AFFINITY AND MATTER

ELEMENTS OF CHEMICAL
PHILOSOPHY 1800–1865

AFFINITY AND MATTER

ELEMENTS OF CHEMICAL PHILOSOPHY 1800–1865

BY

TREVOR H. LEVERE

CLARENDON PRESS · OXFORD

1971

*Oxford University Press, Ely House, London W.*1

GLASGOW   NEW YORK   TORONTO   MELBOURNE   WELLINGTON
CAPE TOWN   SALISBURY   IBADAN   NAIROBI   DAR ES SALAAM   LUSAKA   ADDIS ABABA
BOMBAY   CALCUTTA   MADRAS   KARACHI   LAHORE   DACCA
KUALA LUMPUR   SINGAPORE   HONG KONG   TOKYO

PRINTED IN GREAT BRITAIN
BY BUTLER AND TANNER LTD
FROME AND LONDON

TO GODFREY AND VICKI
LEVERE

PREFACE

HUMPHRY DAVY once attributed to Newton, Hooke, and Boscovich the 'sublime speculation' that a fundamental simplicity underlay the apparent variety of chemical species and qualities. Complicated arrangements and interactions of original matters and powers, few in kind, were responsible for the rich diversity of chemical phenomena. Coleridge and Kant favoured similar lines of conjecture, agreeing on their importance for an understanding of affinity and matter, the transcendental parts of chemistry. Chemists and philosophers, theologians and physicists became involved in the problems of chemical affinity, and shared in their elaboration and solution.

Chemistry is now unquestionably an integral part of physical science, but looking back it seems to have been slow in winning that recognition. Its achievement owes much to the emergence in the nineteenth century of chemical affinity as a unitary concept in chemical theory, linking not only different branches of chemistry, but also chemistry as a whole with physics. Affinity played a crucial role also in the development of the concept of matter, generally significant for the history and philosophy of science, and not limited to the history of chemistry. Although different aspects of the history of affinity have been examined separately, this study is perhaps the first attempt to relate them both to one another and to the intellectual background of the nineteenth century. Related studies exist already for the eighteenth century, in the recent works of R. E. Schofield and A. Thackray.

By the end of the nineteenth century, the unifying concepts of chemical affinity and matter theory had led on to more specific questions about chemical structure, combining ratios, and chemical energetics. Attempts at the solution of these problems came from stereochemistry, valency, and thermodynamics, disciplines not yet established at the beginning of the nineteenth century. This book deals with the role of affinity prior to the age of specialization.

The concept of affinity in its protean shape during the past century is scarcely familiar to historians of science today, except of course to specialists. Survey histories of chemistry now deal only, or mainly, with its more conspicuous aspects. Relying on such works, historians interested mainly in physics or biology do not recognize the importance of affinity as an early link between their fields through chemical theories.

Affinity was variously invoked to account for chemical reaction, selectivity, stability, energetics, and crystallization, and emerged as the underlying explanation of natural chemical classification. Such a *deus ex machina* is not willingly confined by narrow definition. Affinity and matter are accordingly considered here as part of a wide chemical philosophy, so that several developments that had important repercussions for affinity are examined, even when they were not primarily concerned with it.

The book is divided into seven chapters, each being largely self-contained and comprising studies of the work of selected individuals, supported by background bridges and surveys where necessary. Thus a sequence of intellectual biographies forms the backbone of the book; the concept of chemical affinity did not exist independently of those men who contributed to its development. In some cases, personal predilections, religious, social, or philosophical, exerted a demonstrable influence on the formation and development of assumptions about the nature and interrelation of matter and chemical force. These influences have been considered where they were effective, so that the book should be of value to historians of intellectual fields outside 'science', narrowly interpreted. Newton's influence on the eighteenth century, Coleridge's friendship with Humphry Davy, the influence of natural theology on the selection of scientific theories, and scientific controversies revolving around *Naturphilosophie* are among the topics 'external' to science that yet play a part in this story.

The biographical approach has rendered a continuous treatment impossible, and detailed descriptions of experimental procedure are omitted. This study is presented as a sequence of related essays rather than as a complete history of nineteenth-century chemical philosophy.

The table of contents indicates the structure of the book. The bibliography, whilst referring to books profitably consulted, omits detailed references to periodical literature and manuscript material, thereby avoiding excessive length. Use of the index and footnotes to the text of the present work should enable the reader to trace sources not described in the bibliography.

This book is based on a D.Phil. thesis in the University of Oxford, substantially modified for publication. Some of the contents of chapters 2 and 4 were presented at seminars in the University of Toronto in 1968 and 1970, while part of chapter 6 has already appeared in *Ambix* (1970), and is reprinted here by permission of the Editor.

This book owes much to many people. I am particularly grateful to Dr. A. C. Crombie, who supervised the work for my thesis, and to Professor L. P. Williams, who provided stimulating and much-appreciated criticism and discussion. I should like to thank also Dr. N. Fisher, Professor S. Drake, Dr. J. H. Brooke, Dr. M. J. S. Hodge, Dr. D. M. Knight, and Dr. C. A. Russell for much valuable advice, Miss C.

MacBrien, who typed the manuscript, and my wife, who endured while this book took shape. I have relied heavily and gratefully upon the librarians and archivists of several institutions, notably the *Académie des Sciences*, the Bodleian Library, Oxford, the British Museum, Cambridge University Library, the Institution of Electrical Engineers, London, the Royal Institution of Great Britain, the Royal Society of London, and the University of Toronto. I am grateful also to Mrs. C. Colvin for the privilege of examining her collection of Maria Edgeworth's correspondence.

Manuscripts are cited with permission from the following institutions: *Archives de l'Académie des Sciences*; Cambridge University Library (by courtesy of the Syndics of Cambridge University Library); *Institut de France*; the Institution of Electrical Engineers; the Royal Institution of Great Britain; the Royal Society, London (copyright of the material used remains in the hands of the Royal Society); Trinity College Library, Cambridge (by courtesy of the Master and Fellows of Trinity College).

Toronto   T.H.L.
20 June 1970

CONTENTS

LIST OF PLATES

(between pp. 110 *and* 111)

ABBREVIATIONS FOR JOURNALS

Abh. k. Akad. Wiss. Berlin
Abhandlungen der Königlichen Akademie der Wissenschaften zu Berlin.
(Berlin.)

Am. J. Sci Arts
The American Journal of Science and Arts. (ed. Silliman. New Haven.)

Alembic Club Repr.
Alembic Club Reprints. (Edinburgh.)

Ambix
Ambix. (Society for the Study of Alchemy and Early Chemistry, London.)

Ann. Chem. Pharm.
Annalen der Chemie und Pharmacie. (ed. Liebig. Lemgo, Leipzig, Heidelberg.)

Ann. Chim.
Annales de chimie. (Paris.)

Ann. Chim. Phys.
Annales de chimie (afterwards *et de physique*). (Paris.)

Ann. Phys.
Annalen der Physik. (ed. Gilbert. Halle and Leipzig.)

Ann. Phil.
Annals of Philosophy; or Magazine of Chemistry, Mineralogy, Mechanics, Natural History, Agriculture, and the Arts. (ed. T. Thomson. London.)
[C. in *The Philosophical Magazine.*]

Ann. Sci.
Annals of Science. (London.)

Archs Hist. Sci.
Archives Internationales d'Histoire des Sciences. (Paris.)

Archs Inst. gr.-duc. Luxemb.

Archives. Institut grand-ducal de Luxembourg. (Luxembourg.)

The Athenaeum
The Athenaeum. (London.)

Br. J. Hist. Sci.
British Journal for the History of Science. (London.)

Br. J. Phil. Sci.
British Journal for the Philosophy of Science. (Edinburgh.)

Bull. Soc. Chim. Fr. (Docum.)
Bulletin de la Société Chimique de France. (Documentation). (Paris.)

C.r. Acad. Sci., Paris
Compte rendu hebdomadaire des séances de l'Académie des Sciences. (Paris.)

C.r. mens.
Compte rendu (mensuel) des travaux chimiques. (ed. Gerhardt. Paris.)

Centaurus
Centaurus. (Copenhagen.)

Chem. Ann.
Chemische Annalen für die Freunde der Naturlehre. (ed. Crell, Helmstadt.)

Chem. Gaz.
The Chemical Gazette. (London.)

Chem. News, Lond.
The Chemical News and Journal of Physical Science. (London.)

Chem. Tech.
Chemie-Techniek. (Leiden.)

Chymia
Chymia. Annual studies in the History of Chemistry. (Philadelphia.)

Contemp. Phys.
Contemporary Physics. (London.)

Edinb. phil. J.
The Edinburgh Philosophical Journal,
exhibiting a View of the Progress
of Discovery in Natural Philosophy.
(ed. Brewster and Jameson. Edinburgh.)

Edinb. Rev.
The Edinburgh Review, or Critical Jour-
nal. (Edinburgh.)

Encycl. Brit.
The Encyclopaedia Britannica.

G. Fis.
Giornale di fisica, chimica, e storia
naturale. (ed. P. Configliachi and L. G.
Brugnatelli. Pavia.)

Hibbert J.
The Hibbert Journal. (London.)

Hist. Acad. Sci.
Histoire de l'Académie Royale des
Sciences. (Paris.)

Hist. Sci.
History of Science. (Cambridge)

Isis
Isis. (Cambridge, Mass.)

Jahresber. Fortschr. Chem.
Jahresbericht über die Fortschritte der
Chemie. (Braunschweig.)

J. Chem. Phys.
Journal für Chemie und Physik. (ed.
Schweigger. Nürnberg.)

J. chem. Soc.
Journal of the Chemical Society. (London.)

J. Hist. Ideas
Journal of the History of Ideas. (Lan-
caster, Pa.)

J. Pharm.
Journal de pharmacie et des sciences
accessoires. (Paris.)
[C. as *Journal de pharmacie et de*
chimie.]

J. Phys.
Journal de physique, de chimie, et
d'histoire naturelle. (Paris.)

J. R. Instn
Journal of the Royal Institution. (London.)

K. svenska VetenskAkad. Årsb.
Kungliga svenska vetenskapsakademiens
årsbok. (Uppsala and Stockholm.)

K. svenska VetenskAkad. Handl.
Kungliga svenska vetenskapsakade-
miens handlingar. (Uppsala and Stockholm.)

Lychnos
Lychnos. (Uppsala and Stockholm.)

Mém. Acad. Sci.
Mémoires de l'Académie (Royale) des
Sciences. (Paris.)

Mem. chem. Soc.
Memoirs of the Chemical Society. (London.)

Mém. Soc. Arcueil
Mémoires de physique et de chimie de la
Société d'Arcueil. (Paris.)

Neues allg. J. Chem.
Neues allgemeines Journal der Chemie
(ed. Gehlen. Berlin.)

Nicholson's J.
Journal of Natural Philosophy, Chemis-
try, and the Arts. (ed. Nicholson. London.)
[C. in *The Philosophical Magazine.*]

Obs. Phys.
Observations sur la physique, sur l'his-
toire naturelle et sur les arts. (ed. Rozier.
Paris.)

Phil. Mag.

The Philosophical Magazine, or Annals of Chemistry, Mathematics, Astronomy, Natural History, and General Science. (ed. Tilloch. London.)
Then *The London, Edinburgh, and Dublin Philosophical Magazine and Journal of Science.* (London.)

Phil. Trans. R. soc.

Philosophical Transactions of the Royal Society. (London.)

Proc. R. Soc. Edinb.

Proceedings of the Royal Society of Edinburgh. (Edinburgh.)

Q. J. Sci. Arts

The Journal of Science and the Arts. (ed. at R. Instn, London.)
Then *Quarterly Journal of Science, Literature, and Arts.* (London.)

Q. Rev.

The Quarterly Review. (London.)

Q. Rev. chem. Soc.

Quarterly Review. Chemical Society. (London.)

Rep. Br. Ass. Advmt Sci.

Report of the British Association for the Advancement of Science. (London.)

Rép. Chim. pure

Répertoire de chimie pure. (ed. A. Wurtz. Paris.)

Rev. Hist. Sci. Applic.

Revue d'histoire des sciences et de eurs applications. (Paris.)

Rev. sci. ind.

Revue scientifique et industrielle. (ed. Quesneville. Paris.)

Scient. Mem.

Scientific Memoirs selected from the Transactions of Foreign Academies and Learned Societies and from Foreign Journals. (ed. Taylor. London.)

Trans. chem. Soc.

Transactions of the Chemical Society. (London.)

Trans. Faraday Soc.

Transactions of the Faraday Society. (London.)

Trans. R. Soc. Edinb.

Transactions of the Royal Society of Edinburgh. (Edinburgh.)

Univ. Stud. Univ. Neb.

University Studies, University of Nebraska. (Lincoln, Nebraska.)

1

INTRODUCTION

THE history of any intellectual concept may be thought of as a more or less continuous thread, woven into and supported by the multiplicity of related strands that make up the back-cloth of contemporary thought and society. The task of historical research, like that of scientific research, is one of continuously alternating expansion and contraction. In the first stage individual threads are identified and described; in the second, they are combined and woven into a new tapestry, which will become a necessary presupposition for the subsequent identification of new threads and the re-evaluation of old ones. Of course the two stages are complementary, and their complete separation is impossible. Nevertheless, at different periods of historical scholarship, one stage or the other must of necessity predominate.

The history of chemistry has emerged from Whiggishness, which regarded the science as almost consciously progressing towards the superior present, while the stubborn idols of error toppled, one by one, before the invincible knights of scientific advance. This attitude may have arisen partly because many historians had been active combatants,[1] and tended to see all issues with the same wartime simplicity; and of course the 'right' ideas triumphed. Recent historical research, however, has sought a new key to the map of nineteenth-century chemistry, in the development of a small number of significant concepts, such as those of atomism, imponderable fluids, and structure.[2] New patterns are emerging.

The concept of affinity, in the protean shapes it adopted throughout its history, was for long a fundamental concept of theoretical chemistry, providing a matrix for the interrelation of all other chemical concepts, and for the unification of chemistry with physics. It was flexible enough to be useful in the most varied contexts, but its precise meaning was far from universally agreed. A broad definition might be that chemical affinity was that whereby substances entered into or resisted decomposition. Different expansions of this definition could, indeed did, produce widely antithetical interpretations of affinity.

The implicit divisions went beyond the theories themselves to under-

[1] e.g. H. Kopp, A. Wurtz, and, more recently, J. R. Partington.

[2] D. M. Knight, *Atoms and Elements*. London (1967); R. Fox, *The Caloric Theory of Gases from Lavoisier to Regnault*. Oxford (1971); N. Fisher, Ph.D. thesis, University of Wisconsin (1971).

lying attitudes towards science. Some chemists sought models for explanation, while others sought merely formal relations between the items of their scientific knowledge. The former, the subjects of this book, generally regarded affinity as being a quality of matter, arising from the nature of matter, so that their theories of matter comprehended the problem of affinity. The causes of chemical combination were extremely important in their natural philosophy, bearing directly on all the fundamental issues concerning change, substance, motion, and the elements. There were, however, many chemists in the eighteenth and nineteenth centuries who, being primarily concerned with nomenclature and taxonomy, constructed affinity tables from empirical data, using them for classification and prediction while trying to avoid speculative and metaphysical chemistry.[3]

Prelude

The idea of affinity is extremely old. Many attempts have been made at identifying its origins and the first use of the term. Such attempts, except in their most general conception, are surely futile, for affinities lie at the basis of all magic, and thereby antedate science.[4] The concept was born well before any glimmerings of modern chemistry: certainly no educated man would have been puzzled by the term in the Renaissance, whatever he may have understood by it.

Affinities in alchemical and hermetic lore were often portrayed as living attributes of matter. Some references to the affections of bodies and to their quasi-spiritual affinities may have been metaphorical, but the 'secret virtues of sympathy and antipathy' more often bore animist connotations.[5] Francis Bacon, for example, truly a figure of the intellectual watershed, relied consistently more on organic than on mechanical analogies in framing his natural philosophy. Like Aristotle, he believed that an examination of motion and its causes should underlie the study of nature; but he also believed that '[d]ispute and friendship are the spurs to motion in nature, and the keys to her works'.[6]

It is certain that all bodies whatsoever, though they have no sense, yet they have perception; for when one body is applied to another, there is a kind of election to embrace that which is agreeable, and to exclude or expel that which is ingrate; and whether the body be alterant or altered, evermore perception precedeth operation; for else all bodies would be like one to

---

[3] See A. M. Duncan, 'The functions of affinity tables and Lavoisier's list of elements'. *Ambix* **17**, 28–42 (1970).

[4] R. P. Multhauf, *The Origins of Chemistry*, p. 299. London (1966).

[5] F. Bacon, *Works*, vol. 1, p. 197. 2 vols., Ball, London (1838).

[6] ibid., vol. 2, p. 559.

another . . . It is therefore a subject of a very noble enquiry, to enquire of the more subtile perceptions; for it is another key to open nature, as well as the sense; and sometimes better. And besides, it is a principal means of natural divination; for that which in there perceptions appeareth early, in the great effects cometh long after.[7]

Clearly Bacon had recognized the problem of chemical affinities; nor was he alone in this. From the beginning of the seventeenth century, references to the problem abound.[8] At first, tentative solutions were proposed in animist terms—similar substances were assumed to have somehow an attraction or sympathy for one another. Bacon thought that the attraction of gold for mercury was brought about by a sympathetic corporeal emission.[9] Such assumptions underlay the widespread acceptance of sympathetic emanations to account for such diverse phenomena as witchcraft,[10] magnetism,[11] and Sir Kenelm Digby's sympathetic powder.[12] Corpuscularianism did not prevent belief in the emotions of matter.

Nevertheless, animist explanations of affinity did come under attack, especially from the iconoclastic corpuscular philosophy, which struck at so many revered orthodoxies with such sweeping success. Robert Boyle, foremost British chemist among the new philosophers, was convinced that the 'elementary War'[13] was fictitious:

I look upon amity and enmity, as affections of intelligent beings, and I have not yet found it explained by any, how those appetites can be placed in bodies inanimate and devoid of knowledge, or so much as sense. And I elsewhere endeavour to shew, that what is called sympathy and antipathy between such bodies does, in great part, depend upon the actings of our own intellect . . .[14]

Boyle attempted to gain acceptance for chemistry as a reputable and respectable part of the new, mechanistic natural philosophy.[15] He endeavoured to prove the great advantages for chemistry of the corpuscularian approach, claiming that all qualitative differences between

[7] ibid., vol. 1, p. 176.

[8] See, for example, Multhauf, op. cit. (4), p. 302.

[9] M. Hesse, *Forces and Fields*, p. 94. Totowa, New Jersey (1965).

[10] B. Willey, *The Seventeenth-Century Background*, pp. 176–83. Penguin (1962).

[11] W. Gilbert, *De Magnete* (London, 1600), trans. P. F. Mottelay, pp. 97–104. Dover, New York (1958).

[12] K. Digby, *Of the Sympathetick Powder*. London (1669).

[13] cf. J. B. van Helmont, *Oriatrike*, pp. 51–2. London (1662).

[14] R. Boyle, *Works of the Honourable Robert Boyle*, ed. T. Birch, vol. 4, p. 289. 6 vols. (1772).

[15] M. Boas, *Robert Boyle and Seventeenth-Century Chemistry*. Cambridge (1958). L. T. More, *The Life and Works of the Honourable Robert Boyle*. New York (1944) and *J. Hist. Ideas* **2,** 61–76 (1941) gives a broader picture of Boyle, and also deals with his alchemy.

substances could be derived, at least in principle, from the 'Motion, Size, Figure, and Contrivance of their own Parts . . .'[16] There could thus be no place for animism in science. Boyle did give practical recognition to affinities, for example when considering the displacement of ammonia by caustic potash.[17] In explaining such phenomena, however, he remained within the limited and limiting frame of his natural philosophy. 'Those hypotheses do not a little hinder the progress of human knowledge, that introduce morals and politics into the explications of corporeal nature, where all things are indeed transacted according to laws mechanical.'[18] After God, local motion among the particles of bodies was the main secondary cause of all that happened in nature. Geometrical factors then decided which corpuscles were best disposed to come together with others of congruent or complementary pattern. Affinity, clearly, did not appear as a distinct force within the corpuscular philosophy. Boyle's younger contemporary and friend, Isaac Newton, went beyond corpuscularianism to a dynamical atomism involving matter and force. The problems of chemical affinity and other qualities of matter were thereby subjected to a drastically new conceptual approach.

Isaac Newton

Boyle had helped to advance the operational and experimental side of chemistry, but his limiting commitment to mechanism rendered largely sterile his long-term influence on chemical theory.[19] Newton, in contrast, was undoubtedly the seventeenth century's main contributor to the theory of chemical affinity. His ideas, variously modified and interpreted, underlay much chemical philosophy throughout the eighteenth century, and constituted an exceedingly viable tradition, which survived well into the next century.[20]

Paradoxically, Newton's enormous prestige also inhibited heterodox theorization about chemical phenomena.[21] The chemists' very acceptance of the universal validity of gravitation restrained many of them from further speculation about forces in chemistry. Indeed, it removed all need for such speculation—now that they understood the world, they

[16] R. Boyle, *The Origine of Formes and Qualities*, preface. London (1667).
[17] H. Kopp, *Geschichte der Chemie*, vol. 2, p. 294. Leipzig (1931).
[18] Boyle, op. cit. (14), vol. 3, p. 607.
[19] R. Boyle, *Experiments, Notes &c. about the Mechanical Origin or Production of divers particular Qualities*, pp. 19–20. London (1675).
[20] For eighteenth-century Newtonianism in chemistry and matter theory, see R. E. Schofield, *Mechanism and Materialism*. Princeton (1970); A. Thackray, *Matter and Force*. Cambridge, Mass. (1970). For a brief introductory account see A. Thackray, 'Quantified chemistry—the Newtonian dream', in D. S. L. Cardwell, ed. *John Dalton and the Progress of Science*, pp. 92–108. Manchester and New York (1968).
[21] H. Metzger, *Newton, Stahl, Boerhaave et la doctrine chimique*, p. 35. Paris (1930).

could concentrate on the details of experimental chemistry. Analyses, descriptions, and recipes multiplied, while original attempts at theoretical interpretation became rarer.

Newton's ideas, then, were tremendously important, and succeeded in imposing a new orthodoxy. In spite of this, the precise nature of his theory of chemistry, if we assume that he had one, is still a matter for argument. Nothing chemical was amenable to mathematical demonstration, and so, instead of a chemical *Principia*, he left behind him a collection of published and unpublished speculations, queries, and suggestions about the lines of future research, many of them bearing on the problem of affinity.

Newton's ideas have been subjected to a multitude of interpretations from the time of their promulgation until the present. This book, although concerned with the influence of Newton's thought on chemical theory, need not try to prove that any single interpretation is true, or even the best available. All that matters here is the way in which posterity received and understood the Newtonian corpus. Among eighteenth-century scholars, as among modern ones, ether theories contended with force theories.[22] Ether theories invoked imponderable fluids and atomic collisions to account for physical and chemical processes; they may be termed *mechanical*. Force theories, on the other hand, explained the same range of phenomena by appealing to attractive and repulsive forces between atomic centres. The relation between matter and force may have been essential or accidental, but, since active powers were involved, the theories may be called *dynamical*.[23]

Both dynamists and mechanists among Newtonian natural philosophers could find convincing support for the plausibility of their positions. Mechanists could have referred to Newton's second paper on 'Light and Colours' (1675), or to his reply to Boyle, offering an 'explication of qualities'.[24] In the latter, for example, Newton began by explaining attraction and repulsion by the distribution and mechanical pressure of the ether, whose excess pressure on the outward facing parts of bodies drove them violently together, causing them to cohere.

Dynamists too could confidently rely on Newton's writings, for instance, his short paper on the nature of acids. Acidity was caused by short-range forces, of which acids partook in an extreme degree. 'For

[22] R. E. Schofield, *Ambix* **14**, 5 (1967).

[23] This distinction between 'mechanism' and 'dynamism' is very similar to Schofield's between 'materialism' and 'mechanism'. Since Newton regarded gravitational attraction as more than mechanical, I have preferred not to use 'mechanism' when forces are involved. When referring to Schofield's *Mechanism and Materialism*, the reader should remember the different meaning Schofield attaches to 'mechanism'.

[24] Letter of 28 Feb. 1678/9: quoted in I. B. Cohen, ed. *Isaac Newton's Papers and Letters on Natural Philosophy*, pp. 251–2. Cambridge (1958).

whatever doth strongly attract, and is strongly attracted, may be call'd an acid.'[25]

Their major sources, however, were certain 'Queries' to the *Opticks*,[26] notably the thirty-first. I shall quote extensively from the last Query because it really founded a major tradition of Newtonian chemistry, which was to flourish, albeit unevenly, throughout the eighteenth century and later be taken up by Davy and Faraday.

Have not the small Particles of Bodies certain Powers, Virtues, or Forces, by which they *act at a distance* . . . but also *upon one another* for producing a great Part of the Phænomena of Nature? For it's well known, that *Bodies* act *one upon another* by the Attractions of a Gravity, Magnetism, and Electricity; and these Instances shew the Tenor and Course of Nature, and make it not improbable but that there may be more attractive Powers than these. For Nature is very consonant and conformable to her self. How these Attractions may be perform'd, I do not here consider.

Newton habitually claimed that he did not know how attractions were effected, but the italicized references to bodies acting upon one another at a distance would surely have encouraged dynamical theorization.[27] Qualifications may be all too readily overlooked.

. . . The Attractions of Gravity, Magnetism, and Electricity, reach to very sensible distances, and so have been observed by vulgar Eyes, and there may be others which reach to so small distances as hitherto escape Observation; and perhaps electrical Attraction may reach to such small distances, even without being excited by Friction.

He proceeded to detail various chemical examples of such attractive forces in action. For example,

When *Aqua fortis* dissolves Silver and not Gold, and *Aqua regia* dissolves Gold and not Silver, may it not be said that *Aqua fortis* is subtil enough to penetrate Gold as well as Silver, but wants the attractive Force to give it Entrance? For *Aqua regia* is nothing else than *Aqua fortis* mix'd with some Spirit of Salt, or with Sal-armoniac; and even common Salt dissolved in *Aqua fortis*, enables the *Menstruum* to dissolve Gold, though the Salt be a gross Body. When therefore Spirit of Salt precipitates Silver out of *Aqua fortis*, is it not done by attracting and mixing with the *Aqua fortis*, and not attracting, or perhaps repelling Silver? . . . And is it not for want of an attractive virtue between the Parts of Water and Oil, of Quick-silver and Antimony, of Lead and Iron, that these Substances do not mix; and by a

[25] Cohen, op. cit. (24), p. 258.
[26] I. Newton's *Opticks*. 1st edn. London (1704); further edns, containing additional 'Queries', 1717, 1721, 1730; 1730 edn repr. Dover, New York (1952). All quotations are from this reprint.
[27] My italics.

weak Attraction, that Quick-silver and Copper mix difficultly; and from a strong one, that Quick-silver and Tin, Antimony and Iron, Water and Salts, mix readily?

Newton, after listing these and other chemical reactions, sought to explain them. He found corpuscular mechanism inadequate without force.

The Parts of all homogeneal hard Bodies which fully touch one another, stick together very strongly. And for explaining how this may be, some have invented hooked Atoms, which is begging the Question; and others tell us that Bodies are glued together by rest, that is, by an occult Quality, or rather by nothing; and others, that they stick together by conspiring Motions, that is, by relative rest amongst themselves. I had rather infer from their Cohesion, that their Particles attract one another by some Force, which in immediate Contact is exceeding strong, at small distances performs the chymical Operations above-mention'd, and reaches not far from the Particles with any sensible Effect. . . .

. . . There are therefore Agents in Nature able to make the Particles Bodies stick together by very strong Attractions. And it is the Business of experimental Philosophy to find them out

Here was a grand research programme for chemists, bold in scope and depth. Newton went on in conjectural anticipation of their findings, ending with a speculation about bodies and active principles in chemistry.[28]

Now the smallest Particles of Matter may cohere by the strongest Attractions, and compose bigger Particles of weaker Virtue; and many of these may cohere and compose bigger Particles whose Virtue is still weaker, and so on for divers Successions, until the Progression end in the biggest Particles on which the Operations in Chymistry, and the Colours of natural Bodies depend, and which by cohering compose Bodies of a sensible Magnitude. . . .
Since Metals dissolved in Acids attract but a small quantity of the Acid, their attractive Force can reach but to a small distance from them. And as in Algebra, where affirmative Quantities vanish and cease, there negative ones begin; so in Mechanicks, where Attraction ceases, there a repulsive Virtue ought to succeed. And that there is such a Virtue, seems . . . to follow from the Production of Air and Vapour, . . . [whose] vast Contraction and Expansion seems unintelligible, by feigning the Particles of Air to be springy and ramous, or rolled up like Hoops, or by any other means than a repulsive Power. . . .
And thus Nature will be very conformable to her self and very simple, performing all the great Motions of the heavenly Bodies by the Attraction of Gravity which intercedes those Bodies, and almost all the small ones of

[28] See J. E. McGuire, 'Force, Active Principles, and Newton's Invisible Realm'. *Ambix* **15**, 154–208 (1968).

their Particles by some other attractive and repelling Powers which intercede the Particles. . . . It seems to me farther, that these Particles . . . are moved by certain active Principles, such as is that of Gravity, and that which causes Fermentation, and the Cohesion of Bodies. These Principles I consider, not as occult Qualities, supposed to result from the specifick Forms of Things, but as general Laws of Nature, by which the Things themselves are form'd: their Truth appearing to us by Phænomena, though their Causes be not yet discover'd.[29]

Thus chemistry appeared as part of a dynamical natural philosophy, governed by short-range forces between atoms; since nature was very conformable to herself, there was probably a broad analogy between chemical and celestial operations, between atoms and planets, gravitational attraction and chemical affinity. 'Thus almost all the phenomena of nature will depend on the forces of particles, if only it be possible to prove that forces of this kind do exist.'

'If only it be possible'; 'How these Attractions may be perform'd, I do not here consider'; 'you will easily discern, whether in these conjectures there be any degree of probability'—Newton's habitual caution never really deserted him. To Roger Cotes, editor of the second edition of the *Principia*, he wrote, 'I intended to have said much more about the attraction of small particles of bodies, but upon second thoughts I have chosen rather to add but one short Paragraph about that part of Philosophy.'[30] Realizing that he did not fully understand molecular forces, he never made dogmatic assertions about them. On the contrary he seems occasionally to have been deliberately and guardedly ambiguous. That is why convincing cases can be made to show that Newton both was and was not a dynamist. Eighteenth-century mechanism and dynamism have received careful and detailed study from recent historians,[31] one of whom has suggested that the most influential view was probably that of the mechanistic etherial school.[32] However, the dynamical school, which followed up the hints about atoms with central forces, undoubtedly contributed the most to the development of affinity theory.

The legacy of Newton—mechanism and dynamism

Eighteenth-century chemists as a whole failed to understand the complexities and subtleties of Newton's thought, for which they substituted artificial qualitative simplicities; but they thought that they did understand him.

[29] Newton, op. cit. (26), '31st Query', pp. 375–406.
[30] A. R. and M. B. Hall, *Isis* 51, 132 (1960): quotation from letter of 2 March 1712/13.
[31] See note (20) above.   [32] Schofield, op. cit. (22).

In 1704, before Newton had published his 'final' views on chemical affinity, John Freind was giving chemistry lectures at the Old Ashmolean Museum in Oxford.[33] Freind based his interpretation on the 'Newtonianism' of John Keill. They both began their work before the ether hints were fully expounded, and were thus uninfluenced by them.[34] In Keill's opinion,

. . . Besides that attractive force, which retains the bodies of planets and of comets in their own orbits, there is also another power inherent in matter, whereby the single particles, of which it consists, do attract one another. This force increases in more than a two-fold ratio to the increase of distance.[35]

Fitting this force into the framework of Newtonian mechanics, Keill proceeded to explain in principle all the phenomena of molecular physics and chemistry, such as cohesion, crystallization, and solution.

Freind gladly took up and developed this approach. Chemistry for him was not to be a catalogue of ill-understood and incompetently executed experiments; nor was he the man to waste time refuting the blatant errors of his Cartesian and animist opponents. There was, there could be, only one chemical system, which dovetailed neatly into the Newtonian cosmos. The very title of his course, dedicated to Newton, resounds with confidence.[36] Starting with the postulated second inherent power of matter, he proceeded to account for the whole of chemistry in dynamical and therefore sound terms. One wonders how Newton reacted to his disciple's enthusiasm.

There was, however, as Freind admitted, one drawback to the system —it was not adequate for mathematical calculation. Fortunately, this would prove to be only a temporary set-back. Once the right relations and constants in the force equations had been determined, chemistry would achieve mathematical certainty. In the meantime, he disarmingly tells us, ''tis enough for our purpose, if . . . we can point out the way . . . 'Tis possible there may be some things, which the greatest Genius and Industry cannot dive into; but if these can't be reduced to the Laws of Mechanism [i.e. dynamism] we had better confess, that they are out of our reach, than advance Notions and Speculations about 'em, which no ways agree with sound Philosophy.'[37] This optimism was

[33] J. Freind, *Praelectiones Chymicae*. London (1709). idem, *Chymical Lectures; in which almost all the Operations of Chymistry are Reduced to their True Principles, and the Laws of Nature*. London (1712).

[34] Schofield, op. cit. (22), makes this point, and adds '. . . the General or aether Scholium did not appear until the second edition of the *Principia* in 1713 and the aether queries, numbers 17 through 24 of the second English edition of the *Opticks*, not until 1717'.

[35] *Phil. Trans. R. Soc.* **26**, 100 (1708).

[36] See note (33) above.

[37] Freind, *Chymical Lectures; in which almost all the Operations of Chymistry are Reduced to their True Principles, and the Laws of Nature*, pp. 96, 149. London (1712).

quite misplaced. A century later William Whewell scathingly commented that attraction of the gravitational kind

has never . . . been worked out into a system of chemical theory . . . Any such attempt, indeed, could only tend to bring more clearly into view the entire inadequacy of such a mode of explanation. For the leading phenomena of chemistry are all of such a nature that no mechanical combination can serve to express them without an immense accumulation of additional hypotheses.[38]

Unless or until advances are made that enable dynamical explanations to account for chemical reactions, 'we must necessarily consider the power which produces chemical combination as a peculiar principle, a special relation of the elements, not rightly expressed in mechanical [dynamical] terms'.

In the event, explanations of affinity strictly within the framework of eighteenth-century Newtonianism proved unsatisfactory and largely sterile. Freind's missing constants were not forthcoming, so that explanations, except in the most general terms, were impossible; increasingly, mathematical and empirical chemistry seemed irrelevant to one another. But there were mildly heretical deviations from dynamical orthodoxy, different shades of interpretation and emphasis within the unsettled Newtonian corpus, which were to prove immensely fruitful in the development of chemical theory in the nineteenth century.

One such variant interpretation was proposed by John Rowning in an attempt to resolve the problem of reconciling the apparent contact of corpuscles with the changes of state undergone by their aggregates. He suggested that there might be at least concentric spheres of attraction and repulsion surrounding the particles of bodies.

If this were allowed, and we might go on, and suppose the Particles of all Bodies to attract and repel each other alternately at different Distances, perhaps we might be able to solve a great many Phænomena relating to small Bodies, which are now beyond the reach of our Philosophy. However, upon the Supposition of the three Spheres of Attraction and Repulsion just mentioned, nothing is more easy than to see how Solids may be converted into Fluids, and Fluids into Solids . . .[39]

Rowning's ideas had a dynamical content, dealing not only with matter, but also with forces of attraction and repulsion that filled most of the volume of bodies. In the second half of the eighteenth century, Kant

[38] W. Whewell, *Philosophy of the Inductive Sciences*. London (1840). Whewell's use of the word 'mechanical' here indicates that he regarded Newtonian physics as a 'mechanical' system; but since forces are involved, Newtonian 'mechanics' are termed 'dynamical' in this book.

[39] J. Rowning, *A Compendious System of Natural Philosophy*, vol. 2, pp. 5–6. London (1774). A standard textbook that went through several editions—the Bodleian Library has four between 1738 and 1759.

and Boscovich independently took the final step towards a wholly
dynamical chemistry where forces alone constituted matter. In spite of
a broad analogy, there are definite and significant differences between
the theories of matter proposed by Kant and by Boscovich.

Immanuel Kant was a provincial professor, rational and almost
pietist in his outlook. Next to individual liberty, he admired the
analytic, scientific spirit. He began to philosophize convinced of Christianity and of the laws of Newtonian physics, and a major aspect of
his work appears as a rationalization about these two fixed points. His
philosophical position had to be at least compatible with them; at best
it might lend them valuable support. Matter and force, as well as space
and time, accordingly received his most careful consideration. Within
his general system, Kant propounded a theory of matter that was to be
of importance in the development of nineteenth-century ideas about
chemical affinity.

He began by claiming that natural science, properly so-called, could
be derived entirely from *a priori* principles. It was a solecism to regard
as a true science any subject whose laws depended on experience.[40]
Kant was merely defining his terms; however, Nature Philosophers
extended this almost wilfully to signify that what they were accustomed
to regard as science must necessarily be derivable from *a priori* principles—hence much metaphysical confusion.

Chemistry, clearly, could achieve only empirical certainty, and should
thus be regarded as a systematic art rather than as a true science.
Chemical phenomena involved no apparent necessity, and therefore
their explanations could not be scientifically satisfactory—the best that
chemistry could aim at was the assumption of a systematic historical
form.[41] A metaphysical chemistry, firmly embedded in a unified mathematical natural philosophy, was accordingly impossible.

It was, however, possible to give an *a priori* derivation of a theory of
matter and motion. A long and closely reasoned argument led Kant to
conclude that matter, the movable in space,[42] filled space by its moving
forces of attraction and repulsion,[43] which in themselves constituted
matter.[44] The repulsive force was limited in extent for any part of
matter, and formed its boundaries. The attractive force, on the other
hand, 'extends itself directly throughout the universe to infinity, from
every part of the same to every other part. . . . Because the original
attractive force pertains to the essence of matter, it belongs to every
part of the same, to act directly at a distance'.[45]

---

[40] I. Kant, *Metaphysical Foundations of Science* (Riga, 1786), trans. B. Bax, p. 138.
London (1883).   [41] ibid., p. 141.   [42] ibid., pp. 147, 150.   [43] ibid., p. 170.
[44] ibid., p. 172. Different degrees of repulsion produced different specific kinds of
matter.   [45] ibid., p. 191.

Whether or not there were centres about which the forces operated was not made clear, and the ambiguity was underlined in Kant's second antinomy,[46] where he tried to show that the discrete monadic character of matter, together with its indivisibility, could be demonstrated. One can read a concept of point atomism into Kant, notably in his *Monadologia Physica* of 1756—but not with any confidence. It would be more accurate to describe his theory as one of force atomism.[47]

Now although chemistry was not itself a true science, it was nevertheless legitimate to inject into it the relevant conclusions yielded by truly scientific natural philosophy. This is just what Kant did. He first defined chemical effects as those whereby 'matters . . . change the combination of their parts reciprocally by their own forces while at rest'[48] —a strange definition in view of the fact that Kant regarded the whole of natural science as a doctrine of motion. He may, however, have been concerned with potential motion.

Kant speculated on the nature of chemical combination. Was chemical solution a complete mutual penetration of forces, and therefore of matter?[49] Even if mutual penetration was not total, it might be so intimate that no art could subsequently overcome it.

But this is not the place to point out hypotheses for special phenomena, but only the principle according to which they are all to be judged. Everything that relieves us of the necessity of having to recognise empty spaces, is a real gain to natural science. For these give far too much freedom to the imagination, to supply the want of accurate knowledge of nature by fancy. Absolute vacuity and absolute density are, in natural science, much the same as blind chance and blind fate in metaphysical science, namely, stumbling-blocks for the investigating reason, by which either fancy occupies its place, or it is lulled to rest on the pillow of occult qualities.[50]

Kant came down heavily against mechanical corpuscularianism, advocating instead that chemistry be explained in dynamical terms; the great variety of matters could then be accounted for on the basis of originally different repulsive forces.[51]

---

[46] I. Kant, *Critique of Pure Reason*, trans. N. K. Smith, pp. 402–9, A434/B462–A443/B471. Toronto, New York (1965).

[47] See *Encyclopedia of Philosophy*, ed. P. Edwards, vol. 2, pp. 444 ff. 8 vols., New York and London (1967). Point atoms are centres of force or power, whose relations and interactions give rise to the sensible phenomena interpreted as material. Force atoms are here defined simply as units of force or power, rather than as point centres of force or power; otherwise, they may be defined similarly to point atoms. Clearly, both point atomism and force atomism are dynamical, and the latter may include the former. The term 'force atom' conveys less information than 'point atom', and is therefore particularly useful for describing some general and very imprecise notions about matter. I shall avail myself of this valuable imprecision in terminology to avoid saying more than I mean about certain ideas, notably Michael Faraday's.

[48] Kant, op. cit. (46), p. 207.

[49] ibid.   [50] ibid., p. 210.   [51] ibid., p. 212.

Here, for the moment, let us leave Kant, carrying with us the picture of a dynamical system of chemistry elaborating itself within a plenum of forces, and let us turn to Boscovich. Recent historians have given Father Boscovich considerable attention, and have rescued him from undeserved obscurity.[52] The inevitable result has been a temporary imbalance of illumination, favouring Boscovich above other contenders as being the source for nineteenth-century dynamical chemistry and point atomism.[53] One should, however, remember the influence of Kant and his successors on German chemistry—where this was dynamical, it was largely Kantian. Furthermore, one should not ignore the ability of a long-thriving British tradition of force atomism, exemplified by Rowning, to provide explanations for almost all the phenomena whose elucidation has been thought to require Boscovichean atomism. The general type of influence is unmistakable; but precise attribution within the tradition is quite another matter. Boscovich's ideas do, nevertheless, appear in the writings of major nineteenth-century chemists.

Boscovich, like Kant, was motivated by a desire for unity and simplicity in his theories, which led him to describe his main work as a 'Theory of Natural Philosophy deduced from a single law of powers'.[54] He regarded matter as consisting of geometrical points individually without properties, but producing, by their spatial relations, a system of powers or tendencies to motion.[55] Since the powers were in functional dependence on the distances between points, the result 'is not merely an aggregate of "forces" combined haphazard, but . . . it is represented by a single continuous curve, by means of abscissae representing the distances and ordinates representing the "forces" '.[56] Boscovich stressed that the physical world was a continuum, in which 'forces' or powers varied continuously with the distance between particles. This, coupled with his single law of 'force', ruled out separately existing attractive and repulsive forces like those proposed by Kant.

One great advantage in chemistry of explaining matter wholly in terms of force was that this theory could account for the phenomena of *elective* affinity, and for the ensuing specificity of chemical reactions, in a way that the simpler force relations of Newtonian dynamics could not.[57]

---

[52] Boscovich's life and works are described in L. L. Whyte, ed. *Roger Joseph Boscovich.* London (1961).

[53] See, for example, the works of L. Pearce Williams and D. M. Knight.

[54] R. J. Boscovich, *Theoria Philosophiae Naturalis* (Venice, 1763), trans. J. M. Child, p. 13. Chicago (1922). Child, however, regularly translates *vis* by 'force'. Because of Boscovich's special caution (see note 55), 'power' is substituted here for 'force', and elsewhere below inverted commas are used around the word force.

[55] ibid., p. 39.   [56] ibid., p. 13.

[57] L. P. Williams in Whyte, op. cit. (52), pp. 154, 156.

Boscovich's fundamental particles, his point atoms, were immutable. However, when they approached to very small distances, they could form a system in stable equilibrium, thereby producing tenacious 'primary particles'. A similar equilibrated system of interlocking forces among these primary particles could produce less stable particles of a higher order of complexity, which could in turn form even less stable

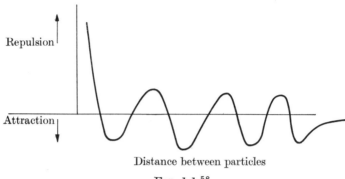

FIG. 1.1.[58]

aggregates. Thus the chemical elements were made up of a complex of interlocking forces.[59] Chemical combination was possible when the force fields of two such complex particles matched, and were in turn capable of interlocking. Boscovich described the situation for systems of two 'atoms' only, but the argument could be extended to more complicated systems.

First of all . . . it may happen that a particle consisting of even two points may attract a third point situated at the same distances from the middle point of the distance between the two points throughout the whole of a certain interval of space, or they may repel it throughout the whole of the same interval, or neither repel or attract it anywhere; in the first case we have a pair of attractions that are equal and in the same direction; in the second case a pair of repulsions that are also equal and in the same direction, and in the third case an attraction and a repulsion that are equal to one another cancelling one another. Also, to a far greater degree, the sum of the forces for the whole of any particle upon the whole of another particle even when situated at this same distance, if the mean for each is considered, will be altogether different from one another for a different distribution of the points. Thus, in one particle attractions will prevail, in another repulsions, and in a third equal and opposite forces will cancel one another. Hence there will be particles acting upon one another with forces that are altogether

[58] Fig. 1.1 (simplified) is from Child, op. cit. (55), p. 42.
[59] Boscovich, op. cit. (54), p. 287.

different, according to the different constitutions of the particles; and there will be particles that are approximately without any action upon one another.[60]

Boscovich's theory was certainly capable of comprehensive, if not detailed, application to chemical problems. It lay within the tradition of force atoms that we have already sketched, and therefore appealed to the minority in Britain who were favourably inclined towards such abstract concepts and regarded a unified natural philosophy as a necessary prerequisite to the study of physical science. The combination of empiricism with immaterialism was a peculiarly British one.[61]

In Joseph Priestley, one finds perhaps the prime exemplar of this combination. He was more than a hugely successful but random experimenter, for a comprehensive natural philosophy paralleled his laboratory investigations.[62] Included in this philosophy was a system of force atomism,[63] whose main advantage and attraction for this heterodox and controversial divine was its approximation of matter and spirit, which narrowed the gap between the things of this world and divine activity, and made the universe more truly a continuum.

I own that, for my part, I feel an inexpressible satisfaction in the idea of that most intimate connection which, on my hypothesis, myself, and everything in which I am concerned, have with the deity. . . . Let the abettors of the common hypothesis says more than this if they can, or any thing different from this, that shall give them more satisfaction.[64]

Priestley clearly inclined towards Boscovichean point atomism, although he was concerned only with its more general implications. In his *Disquisitions on Matter and Spirit* he discussed point atomism, but expressly refused to extend the discussion to the internal structure of matter, 'concerning which we know, indeed, very little, having few data to argue from'.[65] He was merely concerned to establish that attraction and repulsion were not imparted to matter, but were of its very essence. General speculations of this nature are not likely to exert any immediate effect upon laboratory research, nor does Priestley's scientific work appear to have been directly motivated by the details of Boscovich's theory. Priestley's few theoretical interpretations, however, lie within the broad tradition of Newtonian dynamism.[66] Boscovich,

[60] ibid., p. 301.   [61] L. P. Williams in Whyte, op. cit. p. 159.

[62] R. E. Schofield, *Ambix* 14, 1–15 (1967).

[63] J. Priestley, *Disquisitions on Matter and Spirit*. 2nd edn, 2 vols., London (1782). idem, *History and Present State of Discoveries relating to Vision, Light and Colours.* London (1772).

[64] Priestley, *Disquisitions on Matter and Spirit*, vol. 1, p. 43. 2nd edn. 2 vols., London (1782).

[65] ibid., p. 35.   [66] Schofield, op. cit. (62), 10.

although critical of Newton, was largely inspired by the 'Queries' to the *Opticks*, and Priestley, in alluding to the ideas of Newton and of Boscovich, seems not to have been troubled by their occasional incompatibility.

Most of the figures who have already received mention were primarily natural philosophers, for whom chemistry offered challenging broad theoretical problems. Experimental minutiae were definitely less interesting. Their explanations of chemical affinity tended therefore to be couched in abstract terms, barely related to chemical phenomena. There was, however, another line of Newtonians, who were chemists first and philosophers only second. They concentrated on laboratory practice, often accepting universal attraction as the source of chemical force, without trying to modify or to extend the fundamental theory.[67]

Since the appearance in 1718 of Geoffroy's original affinity table,[68] based on displacement reactions, chemists had become aware of the predictive potentiality of such tables. Some substances could replace others in their combinations; if one assumed that the order of replacement was constant, then an empirically determined table would indicate the course of reaction for any pairs of substances in the table. Cartesian disbelief in attraction inhibited theoretical explanations of these supposedly fixed relations. Cartesian explanations, such as those proposed by Lemery in the seventeenth century,[69] had been dropped; but the alternative Newtonian paradigm remained anathema to the *Académie des Sciences* in Paris.

As a result of this rejection of both Newtonian and Cartesian contenders, all theorizing about the cause of affinity was shunned, and attempts were made to describe chemical phenomena in theory-free terms. Tables of affinities could be used for predicting the course of reactions, because they had been shown by experiment to embody empirical generalizations: no theoretical content, however, was implied. The tables were functional, not metaphysical. Thus Geoffroy wrote of relations (*rapports*), instead of affinities or attractions.[70] Similarly Fontenelle, discussing Geoffroy's tables, wondered what would account for them. 'It is here that sympathies and attractions would be altogether relevant, if only they existed.'[71]

Probably the dearth of new affinity tables after Geoffroy's was connected with mistrust of their theoretical foundations. There was one table in 1730; the next one did not appear until 1749. Interest in affinity tables was revived in the 1750s, perhaps stimulated by the

[67] Metzger, op. cit. (21).
[68] Geoffroy, *Mém. Acad. Sci.* 202–12 (1718).
[69] N. Lemery, *Cours de chymie.* Paris (1675), 9th edn (1701).
[70] Geoffroy, op. cit., (68).
[71] *Hist. Acad. Sci.* 35–7 (1718).

publication of Macquer's *Elémens de chymie théorique* (1749),[72] which devoted a whole chapter to the topic.

All the experiments which have been hitherto carried out, and those which are still being daily performed, concur in proving that between different bodies, whether principles or compounds, there is an agreement, relation, affinity or attraction (if you will have it so), which disposes certain bodies to unite with one another, while with others they are unable to contract any union: it is this effect, whatever be its cause, which will help us to give a reason for all the phenomena furnished by chemistry, and to tie them together.[73]

Affinity might remain incomprehensible, but with its aid one could reasonably explain and reduce into a coherent unity all the disparate phenomena of chemistry. Affinity tables were clearly useful. When Voltaire, arch-anglophile, succeeded in his mission as self-appointed apostle of Newtonianism to the French, the stage was set for continental interpretation of the concept of chemical affinity.

Newtonian attraction, by itself, was incapable of accounting for elective affinities.[74] A solution to the problem was proposed almost incidentally by Buffon, who suggested that chemical attraction, besides being inversely proportional to the square of the distance, was also a function of particular shape.[75] 'Shape, which in the case of the heavenly bodies makes little or no difference to the law of mutual interaction, because their distance apart is so great, makes, on the contrary, all the difference when the distance is very small or nil.' Again, '[a]ll matter is reciprocally attracted according to the inverse square of the distance, and this general law only appears to be departed from in particular instances because of the effect of the shape of the constituent particles on each substance; for each shape is involved as an element of distance.' He suggested that by finding the law of attraction for a given substance, such as mercury, one could then find, by trial and error, the shape that gave this result. This led to the possibility of calculating relative affinities, and of returning to the ideal of a mathematical chemistry.[76]

Macquer, in the first edition of his *Dictionnaire*,[77] had stressed that

[72] A. M. Duncan, *Ann. Sci.* **18**, 189–94 (1962).

[73] P. J. Macquer, *Elements of the Theory and Practice of Chymistry*, trans. A. Reid, vol. 1, p. 12. 2 vols., London (1775).

[74] cf. note (57) above.

[75] Buffon, *Histoire Naturelle*, vol. 13, pp. xii–xx (1765).

[76] Different molecular shapes may produce the same attraction, and therefore the anticipated solutions would be ambiguous. This appears not to have been pointed out at the time.

[77] P. J. Macquer, *Dictionnaire de chymie* (2 vols., Paris, 1766), trans. J. Keir, art. *affinity*. 2nd edn, 3 vols., London (1777).

the cause of affinity was unknown, so that dynamical interpretations were invalid. He was, however, swiftly conquered by Buffon's seductive explanation. In a footnote to the article on affinity in the 1778 edition of his *Dictionnaire* he publicly recanted:

I do not think that I can express in a more clear and precise manner that there are no special, separate laws for affinity; on the contrary, its effects are determined by one, and only one, law, the most comprehensive and the most general law of Nature that has hitherto been formulated; I mean that law according to which every particle of matter attracts every other particle . . .[78]

Bergman too considered Buffon's interpretation highly plausible and buttressed it with independent evidence, adduced from crystallography, that the particles of bodies did have definite and unvarying shapes.[79] Since, however, one could not determine the figure and position of each particle, one was forced back upon numerous experiments to determine the relative affinities between bodies for each individual case.[80] The results of innumerable experiments convinced Bergman that a constant order of affinities prevailed. The consequences for chemistry were momentous.

Will [this order] not, when one ascertained by experience, serve as a key to unlock the innermost sanctuaries of nature, and to solve the most difficult problems, whether analytical or synthetical? I maintain, therefore, not only that the doctrine deserves to be cultivated, but that the whole of chemistry rests upon it, as upon a solid foundation; at least if we wish to have the science in a rational form, and that each circumstance of its operations should be clearly and justly explained.[81]

He recognized that heat could upset and even completely reverse the order of affinities, but regarded this as an external factor operating by means of differential volatilities; it did not change the true order. Some reactions, however, would not occur without heat. Bergman therefore proposed two tables, one for reaction in solution (the wet way) and the other for reaction brought about by heat (the dry way).[82]

Other apparent exceptions to the true order could, he observed, arise from double attraction, where more than two principles acted and the sums of affinities were involved;[83] from a successive change of substances, for example when phlogiston was expelled in the course of reaction;[84] from solubility, which conceals decompositions;[85] and from

[78] Macquer, *Dictionnaire de chymie* art. *affinité*. 2nd edn, 4 vols., Paris (1778).
[79] T. O. Bergman, *Physical and Chemical Essays*, trans. E. Cullen, vol. 2, pp. 11–15. 3 vols., London (1784–91).
[80] T. O. Bergman, *A Dissertation on Elective Attractions*, trans. T. Beddoes, p. 3. London (1785).   [81] ibid., p. 9.   [82] ibid., pp. 14–16.   [83] ibid., p. 18.
[84] ibid., p. 29.   [85] ibid., p. 37.

the effects of an excess of one or other of the reagents.[86] This was a qualitative anticipation of mass action—though not the first. Boyle had recognized the effect of quantity on the course of a reaction.

If one took into account the above sources of error, then, in a series of n substances, one could determine the order of elective attractions by a series of $n(n-1)$ experiments on displacement, precipitation, or exchange reactions.[87] The inadequacy of existing tables of elective attractions was scarcely surprising, since even the 'brief sketch' that Bergman proposed required more than thirty thousand exact experiments! This was indeed a massive programme—if only chemists would carry it out.

Understandably, no one was willing to tackle so vast a project unaided; but tables of affinities, the 'consequences of an improved state of chemistry',[88] began to appear regularly in text-books and lectures. The second edition of the *Encyclopaedia Britannica* (1778) considered that theories of affinity were merely conjectural, 'neither is it a matter of any consequence to a chemist whether they are right or wrong';[89] but the practical importance of the concept was stressed, and the table used by Joseph Black in his chemistry course was published in the article on chemistry.[90]

Black wished to exclude physics from chemistry,[91] while simultaneously admitting that the latter was 'a speculative and philosophical science, proceeding, like all other such sciences, on the relation of cause and effect'.[92] On the speculative side, he admitted that the 'Queries' to the *Opticks* did account for many of the most remarkable chemical phenomena. However, the new galvanic discovery 'seems to derange all the notions of the immediate agent in chemical union or separation'.[93] In spite of his respect for Newtonian doctrine, Black was unwilling to commit himself to belief in the identity between chemical and gravitational attraction; the two 'forces' were clearly analogous, but one could not deduce a firm theory from analogy.

I apprehend that chemical science is . . . obstructed by speculations about the principles of affinity, and, particularly, by the attempts of ingenious men to explain the chemical operations by attractions and repulsions. . . .

[86] ibid., pp. 47 ff., 61 ff.   [87] ibid., pp. 64 ff.
[88] *Encycl. Brit.*, vol. 3, p. 1816. 2nd edn, Edinburgh (1778).
[89] ibid., p. 1808.   [90] ibid., pp. 1816–17.
[91] This was because he considered that the state of contemporary physical theory was insufficiently grounded and advanced for it to lead to explanation in chemistry: however, it was legitimate to prosecute the study of matter from both chemical and physical standpoints (cf. his work on heat), in hopeful anticipation of an eventual synthesis. Meanwhile, chemistry simply could not be reduced to physics, and attempts at this reduction were misguided, both for the reason just given, and because they diverted the chemist from his proper study—that of the chemical changes of matter.
[92] J. Black, *Lectures on Chemistry*, ed. J. Robison, vol. 1, pp. 1 ff., 16. 2 vols., Edinburgh (1803).   [93] Ibid., vol. 1, p. 540.

I may venture to say that no man ever got a clear and really explicatory
notion of a chemical combination by the help of attractions. . . . The chemist
will do more wisely, if he forego all speculations about the ultimate internal
action, and direct his whole attention to the external phenomena . . . [A]ll
the mathematicians in Europe are not qualified to explain a single combination by . . . means [of atomic motions, attraction and repulsion] . . .
Let chemical affinity be received as a first principle, which we cannot
explain; any more than Newton could explain gravitation, and let us defer
accounting for the laws of affinity, till we shall have established such a
body of doctrine as he has established concerning the laws of gravitation.[94]

Nevertheless, when discussing affinities operating in specified reactions, Black used an attractionist vocabulary, and even wondered
whether there might not also be repulsions operating.[95] Cartesian
objections to speaking of chemical attraction were 'certainly unreasonable'.[96]

Black seems to have accepted tacitly some sort of dynamical explanation, while warning that time should not be wasted on such
speculations. The solution of this paradox may be found in a consideration of Black's natural philosophy against the general philosophic background of his period.[97] Hume's scepticism had been in its
turn followed by the common-sense philosophy of Thomas Reid, who
appealed to experience for direct knowledge of a permanent world.
Reid was intensely interested in Black's chemistry, and indeed regarded
Black as 'an instructor and a guide'. For both of them, dynamical
atomism, as accepted by Black's predecessor Cullen, had been transferred from the sphere of physical reality to that of mere hypothesis.
Black was therefore obliged in his philosophy to exclude all questions
about real matter, and to consider only change. Dynamical physics had
lost its logical supremacy. Under these circumstances, Black could
legitimately speculate on dynamism in chemistry; but at the same time
he had to recognize that speculations must be rigorously excluded from
strict science.

Lavoisier was in a comparable position. Although aware of the
potential utility of affinity tables, he declined to discuss the theory of
affinity, the 'transcendental' part of chemistry, in his *Elements of
Chemistry*, because, lacking adequate data for its definition, it remained metaphysical.[98] But perhaps increased experimental knowledge

[94] *Lectures on Chemistry.*, pp. 282–4.
[95] J. Black, 'Experiments upon magnesia alba'. *Alembic Club Repr.* no. 1, pp. 44–5.
(1755).
[96] Black, op. cit. (92), vol. 1, p. 266.
[97] A. L. Donovan, 'Cullen and Black': paper read at Br. Soc. Hist. Sci. meeting,
5 Jan. 1968.
[98] A. L. Lavoisier, *Elements of Chemistry*, trans. R. Kerr, p. xix. Edinburgh (1790).

would one day reverse this situation, and 'the mathematician in his study would be able to calculate any phenomenon of chemical combination in the same way, so to speak, as he calculates the movement of the heavenly bodies'.[99]

Many of his contemporaries assumed that such optimism was justifiable. In 1785 Bryan Higgins told his audience that attraction and repulsion constituted the true difference of bodies;[100] Fourcroy was convinced of the identity of gravitational and chemical forces;[101] Pearson looked forward to a mathematical chemistry;[102] and Guyton de Morveau predicted the computation of equilibrium forces in chemical reaction. Examples could readily be multiplied.[103]

The variants of Newtonianism considered above had important consequences for nineteenth-century chemical theory.

Mechanism and dynamism, however, by no means exhausted the range of eighteenth-century ideas about affinity. Many chemists regarded physical science as at most a partial contributor to the total explanation of affinity; at the far end of the spectrum, Newtonian natural philosophy was denied all relevance to chemical processes. Boerhaave, for example, regarded Newtonianism as compatible with chemical science, but came to view it as being of increasingly little use for an understanding of chemistry.[104] Stahl went further, objecting to Newtonian explanations of chemical phenomena as superficial and totally irrelevant to experimental chemistry. Organic analogies feature in Stahl's writings with more immediate significance than the dead analogies of dynamism or mechanism.[105]

In the preceding pages we have encountered the principal historical antecedents of nineteenth-century affinity theory. The animist thread, encountered in Bacon, runs, often concealed, throughout much of the eighteenth century, to enjoy a brief but spectacular efflorescence among Nature Philosophers in the early nineteenth century.

More significant was the Newtonian addition of forces, however interpreted, to the corpuscularianism of Boyle and the Cartesians. From this arose two closely related traditions. The chemical approach accepted Newtonian attraction, inversely proportional to the square of the distance, as given, and was concerned only with the extension of purely

---

[99] *Mem. Acad. Sci.* 534–5 (1782): quoted in transl by M. Crosland, *The Society of Arcueil*, p. 237. London (1967).

[100] Royal Institution MS. iv, 22e. cf. Duncan, *Ann. Sci.* **23**, 168 (1967). There is a polarity in all matter, and only one kind of attraction. Gravitational attraction is the sum of individual atomic attractions.

[101] A. F. Fourcroy, *Elements of Chemistry and Natural History, to which is prefixed the Philosophy of Chemistry*, trans. R. Heron, vol. 1, p. 36. 4 vols., London (1796).

[102] *Nicholson's J.* **1**, 355 (1797).

[103] *Nicholson's J.* **1**, 110 (1797).

[104] Metzger, op. cit. (21), part 3, *passim.*

[105] ibid., part 2, *passim.*

chemical doctrine. Thus, attraction as a force in chemistry gradually receded in importance, until with Black we find that it has lost its logical primacy, and that chemistry is once again an essentially qualitative science. Some among the chemists, however, and fewer among the physicists and natural philosophers, continued to stress the importance of gravitational attraction in chemical phenomena. In France, encroaching Newtonianism dispelled Cartesian antipathies, and opened the way for Buffon's suggestions about the influence of atomic shape, thereby giving a new lease of life to Newtonian chemistry. In the following century the *Académie des Sciences* clung resolutely to what had generally become an outmoded paradigm.

Besides this literal Newtonianism of the chemists, there was a more physical approach advocated by natural philosophers. Particularly in England, this led to an increasing emphasis on forces; but the culmination of this trend occurred abroad, with the systems of Kant and of Boscovich, where matter became identified with force alone.

We are now ready to enter on a more detailed study of nineteenth-century developments in affinity theory.[106]

[106] I have given but scant attention to many eighteenth-century writers on affinity, who were regarded by their contemporaries as authorities on this topic. Thus Kirwan, whose notions are generously spread through succeeding editions of the *Encyclopaedia Britannica*, has not even been mentioned. The reason is simple; his supposed measurements of affinity were in fact measurements of equivalents, and although celebrated in their time, had almost no effect on subsequent developments. This has been one criterion for exclusion.

In general, those who merely used, measured, or referred to affinities without contributing to the theory have likewise been ignored.

2

HUMPHRY DAVY: ROMANTIC AND DYNAMICAL CHEMISTRY; ELECTRICAL IDEAS ABOUT AFFINITY

DAVY'S was a rare and complex character, harmonizing what in other men would have remained mere discords. At times impulsive and romantic, he was yet capable of conducting the most meticulously ordered investigations.[1] He was lonely and imaginative, and this combination resulted in countless dialogues scattered through his notebooks, where essays on materialism alternate with thoughts on the immortality of the soul, forces jostle with fluids, atomism with dynamism. Davy wrote out the arguments in support of different positions in religious, philosophical, and scientific controversies so reasonably, and with so open a mind, as frequently to defeat conjecture about his opinions.[2] To add to the difficulties, his papers were as carelessly ordered and as readily sacrificed as his ideas. His brother John tells us that 'Letters and papers he very seldom arranged, and his rooms were littered with them. Occasionally they were collected and thrown together in a large cupboard. I remember once his commissioning me to look over this great collection, and to burn such as appeared of no interest.'[3] Happily, John so worshipped his brother that everything possible was rescued, and this wealth provides a basis for the rediscovery of the development of Davy's thought.

Humphry Davy's intellectual aspirations knew no limit; in 1795 he listed the subjects he would study and master:[4]

1. Theology
2. Geography
3. My Profession
   (i) Botany
   (ii) Pharmacology
   (iii) Nosology
   (iv) Anatomy
   (v) Surgery
   (vi) Chemistry
4. Logic
5. Language
   (i) English
   (ii) French
   (iii) Latin
   (iv) Greek
   (v) Italian
   (vi) Spanish
   (vii) Hebrew
6. Physics
7. Mechanics
8. Rhetoric and Oratory
9. History and Chronology
10. Mathematics

[1] cf. D. M. Knight, *Ambix* **14**, 181 (1967).
[2] cf. R. Siegfried, Royal Institution discourse, 3 May 1968.
[3] J. Davy, *Memoirs of the Life of Sir Humphry Davy*, vol. 1, p. 260. 2 vols., London (1836).
[4] Sir Harold Hartley, *Humphry Davy*, p. 12. London (1966).

That he failed to become quite the polymath he intended goes without saying; what is significant is the astonishing sweep of his plans, and the self-confidence that this implies. Davy could and would tackle anything, give it his own imprint, and absorb it into his own eclectic philosophy.

In this chapter we shall examine Davy's contributions to the meaning and growth of the concept of chemical affinity. But an intellect at once so complex and so unified as Davy's makes it impossible to do this by simply pulling out the relevant thread and studying it in isolation. Instead, Davy's overall development must be retained as a back-cloth to the development of our theme. His ideas on chemical affinity must be presented in terms of intellectual biography.

The argument, though simple in essence, is complicated in detail. A summary outline of conclusions reached may help in following the separate sections below.

Davy began his study of chemistry by reading Lavoisier's *Traité élémentaire de chimie*, and intuitively realized that the main weaknesses of Lavoisier's system lay in his list of elements: caloric and oxygen required special attention, the caustic alkalis were probably compound, as Lavoisier suspected, and so, by analogy, were the alkaline earths, which Lavoisier regarded as elements. From the very beginning, Davy was convinced that matter was fundamentally simple, and that it was the task of chemistry to elucidate the arrangements and powers, i.e. the affinities, of the true elements, once these had been discovered. Thus the problem of affinity was fundamental to chemical science: Davy went further, and called it the sublime part of chemistry.

His ideas on matter were Newtonian, well within the dynamical tradition of eighteenth-century Newtonianism: Davy himself rightly saw Boscovichean point atomism as a part of this tradition, and after 1812 he used it in his work. Use does not imply acceptance, and at many other times in the same period he used more corpuscularian forms of Newtonianism.

At the basis of his philosophy of science lay eighteenth-century natural theology: it was to this that Davy owed his seminal convictions of the simplicity, order, unity, and purposefulness of the cosmos; and his theory of matter, together with his life's work in the laboratory, can be seen as attempts to illustrate these convictions by discovery in the natural world.[5] Davy's natural theology is the link between his science and his final work, the *Consolations in Travel*.

Immediately after Volta's discovery of the 'pile' or battery, and its use by Nicholson and Carlisle to decompose water, Davy realized the

[5] See also T. H. Levere, 'Faraday, Matter and Natural Theology'. *Br. J. Hist. Sci.* **4**, 96–107 (1968).

potential of this instrument for overcoming the affinities of bodies, and for trying to reveal their fundamental simplicity of constitution. From the galvanic experiments that followed this realization, there grew, quite naturally, a conviction that chemical affinity and electrical attraction were different manifestations of a single power. Furthermore, Davy considered that different electrical states of matter might account for its different forms; the electrical and chemical powers of matter, displayed by masses and by particles respectively, might account not only for chemical reaction, but also for chemical diversity. Davy believed that this was true, but he was only prepared to say that it was probably true; nevertheless, the idea that chemical and electrical powers of matter were intimately related was, for him, a fact of experience, and one that guided his work for more than twenty years.

The Newtonian conviction that different arrangements resulted in different properties was interpreted by Davy to mean that mechanical arrangement and electrical state were somehow coterminous, and it is this interpretation that accounts for his conclusion in 1814 that diamond differed from charcoal in crystalline form and in conductivity. Davy's discovery of the halogens, and indeed all of his work between the years 1809 and 1814, was connected by researches on the nature of the diamond. At a more fundamental level, however, the search for simplicity and unity provides the true connecting strand that runs through all his work. Physical and chemical properties will ultimately be found to be one, different forms of matter are interconvertible, chemical affinity is implied in a dynamical theory of matter, and chemistry will ultimately be found to conform to simple mathematical laws, comparable to those of celestial mechanics.

All these ideas are related by, and logically derived from, Davy's natural theology. To explain Davy's ideas on affinity (or indeed on anything else) it is necessary to remember their context of Newtonianism and Christianity, rather than the existence of German Idealistic metaphysics.

Romantic and dynamical chemistry

In a notebook of about 1795, Davy enters the study of science as a corpuscularian philosopher in fine seventeenth-century style. 'Motion', he affirms, 'if ever so artfully distributed, it is plain, can produce nothing but motion.'[6] By the following year he had become acquainted with the 'Scotch metaphysicians', suffered his first superficial encounter with Kant, and begun to grapple with the problem of the nature of matter.[7] This problem was to loom large throughout Davy's

---

[6] J. Davy, op. cit. (3), vol. 1, p. 26.
[7] ibid., pp. 36–8.

life, and as early as 1796 he speculated in favour of a dynamical theory of matter: 'Far from being conscious of the existence of matter, we are only conscious of the active powers of some being. By discovering the ratio between the attraction and repulsion of external things and our organs, we should discover philosophy.'[8]

In 1797 he began the serious study of natural philosophy, and read his first major chemical texts,[9] Lavoisier's *Traité élémentaire de chimie*[10] and Nicholson's *Dictionary of Chemistry*,[11] at the age of 18. He intuitively seized on weaknesses in the former's oxygen-centred theory, and went beyond Lavoisier's hints in seeing the vulnerability of his table of elements.[12] He also learned from Lavoisier that theories of affinity were inadequately founded upon experiment, and were thus metaphysical, constituting the transcendental part of chemistry;[13] surely this philosophically forbidden fruit would have excited Davy's curiosity. Theories of affinity and of chemical elements were henceforth to direct the bulk of Davy's experimental and speculative science; it would seem reasonable to suppose that Lavoisier's seminal treatise was a source for both these lines of investigation.

Nicholson's *Dictionary* served only to underline these problems, stating that the supposed chemical elements 'are simple only with respect to our power of decomposing them'; he also hinted that the forces associated with chemical affinity were polar.[14] Davy's later investigations developed this suggestion.

The importance of these two works in directing Davy's enquiries is strongly suggested by a notebook of 1799:

The perfection of Chemical Philosophy or the laws of corpuscular motion must depend on the knowledge of all the simple substances, their mutual attractions and the ratio in which these attractions increase or diminish on increase or diminution of temperature. These being ascertained Chemistry would become a science so far generalised as to enable calculations to be formed with regard to the result of any near approach of particles.

The first step towards these Laws will be the decomposition of those bodies which are at present undecompounded and probably no class of substances will throw greater light on this than the Ammoniacal salts.[15]

Much of the remainder of this notebook is devoted to experiments

[8] J. Davy, *Memoirs of the Life of Sir Humphry Davy*, vol. 1, p. 38.
[9] Hartley, op. cit. (4), p. 12.
[10] A. L. Lavoisier, *Traité élémentaire de chimie*. 2 vols., Paris (1789).
[11] W. Nicholson, *A Dictionary of Chemistry*. 2 vols., London (1795).
[12] For a detailed discussion of Davy and the chemical elements, see D. M. Knight D.Phil. thesis, Oxford University (1964).
[13] Lavoisier, op. cit. (10), vol. 1.
[14] Nicholson, op. cit. (11), vol. 1, p. 154, art. *attraction*.   [15] R. Instn MS. 20b.

on the composition, analysis, and decomposition by heat of the ammoni-acal salts—Davy was always prompt in subjecting his speculations to experimental test, hence, frequently, their short lives and rapid succession.

At this period, Davy's theory of matter and of affinity was one of forces or powers associated with corpuscles, firmly based in the eighteenth-century British tradition exemplified by Rowning.[16] Davy would have called himself a Newtonian, and his views on short-range forces and the nature of matter stemmed directly from a perusal of Newton's '31st Query'. He had studied this query, copied it piecemeal into various notebooks, and accorded Newton the ultimate accolade by putting him (in a fit of overweening egotism and ambition while under the influence of laughing gas) on a par with—Davy!

Newtonian corpuscularianism permeates his *Essay on Heat and Light*.[17] This juvenile effort, crammed with speculations, pleased his mentor and employer Dr. Thomas Beddoes; but it scandalized re-viewers, who savaged it.[18] Davy, much chastened, first moderated and then renounced his claims, and resolved that his future scientific pub-lications would deal with truly experimental chemical philosophy.[19] This rash early publication is invaluable in its revelation of Davy's fundamental tenets in natural philosophy. It gives succinct expression to his conviction that nature is ultimately one, a unified and simple whole resulting from God's comprehensive plan of creation. If there was any single connecting thought running through all Davy's work, it was this. It was quite in accord with the continuing attempts of Christian theologians to establish—against polytheists—that there is one and only one God. Paley had pointed out that if all the parts of nature showed simple similarities one could argue from the unity of nature to the existence of a single Creator; the unity of nature and the unity of God were implicit in one another. Neo-Platonists, with a more animistic view of nature, argued the whole of creation was one living body with one world-soul, and that this unity necessitated the unity of all motions. From this position, as from the more orthodox Christian one, it was but a short step to the conviction that all forces in nature were ultimately one and interconvertible, subject to a single law. This indeed was Davy's conviction, and he saw the different manifestations of the unity of force as governed by 'an energy of mutation impressed by the will of the Deity'.[20]

[16] See p. 10 above.
[17] *The Collected Works of Sir Humphry Davy, Bart.*, ed. J. Davy, vol. 2, pp. 5–86. 9 vols., London (1839).
[18] A. Treneer, *The Mercurial Chemist*, p. 37. London (1963).
[19] R. Instn MS. 20a; J. Davy, op. cit. (3), vol. 1, pp. 81–2.
[20] H. Davy, op. cit. (17), vol. 2, p. 85.

'The modern chemist', said Davy in 1809 (and who was more modern than himself?) '. . . admires her [Nature's] harmony and order even when the causes are above his comprehension. And he is most happy, when in the obscurity of sensible things, he catches a gleam of that intellectual light which'—does what? Davy wasn't sure. He first wrote down 'governs and directs the universe'. He then had second thoughts, scratched out 'directs', and replaced it with 'animates',[21] veering from orthodoxy dangerously towards theosophy, and echoing many a seventeenth-century debate. Coleridge was alarmed by this, and it was perhaps at his instigation[22] that intelligent direction came to replace animation in Davy's cosmos. Either, however, would have guaranteed the fundamental unity of natural forces and natural laws.

The twin factors of Newtonianism and natural theology, clearly accepted at the outset by Davy, made of the natural world a unity where biological and chemical changes were ultimately due to the same divinely impressed laws. At a lower level, the laws of chemistry could illuminate the laws of life:

Hence the knowledge of sublime chemistry or the classification of the attractions and corpuscular motions producing the phenomena of the external world will [not only] be interesting to man as enlarging his ideas and giving grandeur to his conceptions and providing for many of his wants; but as opening the field for discoveries still more important and sublime—the knowledge of the laws of his own existence.[23]

Thus the laws of chemical affinity would become ennobled in this divinely integrated universe. Yet precisely because of his organic grasp of the natural world as a unity, Davy was able both to focus his attention on the minutiae of laboratory chemistry, and to discuss affinity in particular cases.

On the basis of a number of experiments on acid–base combination and on the reactions of non-metals such as sulphur and phosphorus with the metallic oxides, and of speculations (broadly confirmed by experiment) on the affinities that pairs of elements have for one another, he arrived at a number of probable inductions. The first of these was that when two principles combined, their simple attractions were not destroyed, 'but only *modified* by their new state of existence'.[24] Modification of the powers of matter by combination belonged to the force tradition stemming from Newton, and remained consistent with the broad trend of Davy's theory of matter throughout his career.

[21] R. Instn MS. : introductory lecture on electrochemical science for 1809 series.
[22] E. L. Griggs, ed. *Collected letters of Samuel Taylor Coleridge* (4 vols., Oxford, 1956–9): Letter to Davy, 30 Jan. 1809.
[23] R. Instn MS. 1.
[24] R. Instn MS. 20a.

Now the supposition of active powers common to all matter, from whose different modifications all material changes resulted,[25] was consistent also with the dynamism of German Idealism; a pure dynamism equating matter with force. Davy, in the early years of his career retained the conception of actual central corpuscles whence forces radiate. So long as he retained these corpuscles, he remained within the British tradition of force atomism, that is, the tradition of eighteenth-century Newtonianism. But Davy's theory of matter, like all his theories, was there to be used rather than defended, and hence central corpuscles were frequently supplanted by all-pervasive forces without material centres. It was at such times that Davy's ideas were most nearly akin to those of the *Naturphilosophen*. Even then, however, the two views were not identical, despite the fact that Davy's later results were used by Nature Philosophers to illustrate their dynamical schemes.[26]

There is no evidence that Davy ever so much as looked into, let alone gave serious study to, the manifold doctrines of *Naturphilosophie*. Schelling and Fichte may be dismissed from this part of the enquiry. Immanuel Kant, however, remains.

John Davy tells us that his brother studied Kant in 1796.[27] The first coherent English exposition of Kant's philosophy did indeed appear in 1796.[28] If John's date is correct (and there is no particular reason to doubt it), then Humphry probably read *A General and Introductory View of Professor Kant's Principles concerning Man, the World and the Deity* by F. A. Nitsch, would-be apostle for the critical philosophy. There was much in the book to appeal to the youthful Davy. Nitsch defined natural philosophy in terms of sensible phenomena[29]—an important point for empirical scientists—and echoed some thoughts of Newton and Paley, of Cudworth and More:

Natural Philosophy . . . proves with incontrovertible evidence, that all nature consists of one uninterrupted and complex chain of causes and effects, and puts us in mind, that as a chain must have a first link, so the causes in nature must depend upon a first cause. This Science goes still farther, and teaches mankind in very plain and convincing language that the first cause is an intelligent, mighty, omnipresent, wise, and beneficent Being. . . .

[25] J. A. Paris, *The Life of Sir Humphrey Davy, Bart.*, vol. 1, p. 80. (2 vols., London, 1831): H. Davy to Davies Giddy, 10 April 1799.

[26] See Chapter 4 below.

[27] J. Davy, op. cit. (3), vol. 1, p. 37.

[28] F. A. Nitsch, *A General and Introductory View of Professor Kant's Principles concerning Man, the World and the Deity*, London (1796). For an account of the fate of Kant's teaching, see R. Wellek, *Immanuel Kant in England*, 1793–1838. Princeton (1931).

[29] Nitsch, op. cit. (44), p. 9.

Natural Philosophy not only investigates the properties and causes of things, but it also . . . informs us that all the innumerable objects in the universe are but parts of a grand and beautiful whole.[30]

This was very fine—and very general. Nitsch wished merely to persuade his readers to investigate Kant's writings for themselves—impossible for Davy, if, as seems the case, he knew no German[31]—and, for the sake of brevity, Nitsch omitted from his account those of Kant's principles that were immediately concerned with mechanical and experimental science. Yet 1796 did see Davy speculating about dynamical and polar theories of matter.[32] At this date, the evidence, maddeningly, goes no farther.

In 1799, Davy was ensconced in Beddoes' Pneumatic Institution at Clifton; and Beddoes, ever athirst for philosophical novelty, and with a fluent reading knowledge of German, was looking into the possibilities offered by Kant's system.[33] It was at Clifton, too, that Davy met Coleridge. Coleridge was an idealistic dynamist,[34] and he and Davy were powerfully attracted to one another as soon as they met. Intellectually, they enjoyed a symbiotic relationship. Coleridge's enthusiasms were infectious. He could scarcely have refrained from telling Davy about Kant; did Davy imbibe some of his ardour for German philosophy? We are not told; fanciful Davy left no philosophical autobiography. Yet it will be worth balancing probabilities and playing with the pieces of the jigsaw in an attempt to gain a little more insight into the mind of Humphry Davy.

Davy and Coleridge set fire to one another's intellectual ambitions—missionary notes on chemistry and metaphysics flew back and forth between them. Coleridge wildly proposed to begin the study of chemistry,[35] and cheered Davy on his triumphal path to fame:

Work hard, and if Success do not dance up like the bubbles in the Salt (with the Spirit Lamp under it) may the Devil and his Dam take success! . . . Davy! I *ake* for you to be with us.[36]

Chemistry united the opposite advantages of immaterializing the mind without destroying the definiteness of our ideas—nay even while it gave clearness

[30] Nitsch, op. cit. (44), pp. 11, 13.

[31] Davy's plans for his education included the study of seven languages; German was not one of them. I have found absolutely no indication anywhere in Davy's writings to suggest that he could read German. This, of course, does not prove his ignorance. (Hartley, op. cit. (4), p. 12.)   [32] See note (8) above.

[33] C. A. Weber, *Bristols Bedeutung für die englische Romantik und die deutsch–englischen Beziehungen*, pp. 93, 106 ff., 168. Halle (1935).

[34] See, for example, D. M. Knight, *Ambix* **14**, 183 (1967): quotation from *The Friend*, vol. 1, p. 155 n. London (1818).

[35] Griggs, op. cit. (22): Coleridge to Davy, 3 Feb. 1801.

[36] ibid., vol. 1, p. 611: letter of 25 July 1800.

to them—And else that being necessarily joined with the passion of Hope, it was poetical . . . the *Poet* is the greatest possible character . . . Modest creatures![37]

Coleridge and Davy walked together, dreamed together, and above all, talked—talked of poetry and religion, of morals and manners, of chemistry and philosophy. Coleridge, in 1799, was newly returned from Germany, and brimming over with enthusiasm for the Romantic philosophy of nature that he had 'discovered' there. His encounter with Davy, and their ensuing friendship,[38] may thus serve to introduce an interlude in which to sketch the fundamental tenets of post-Kantian idealistic dynamism, and to distinguish them from their Kantian and Newtonian analogues.[39]

Kant warned against the application of *a priori* methods to the explanation of chemical phenomena. Post-Kantian Idealists largely ignored these warnings. This was because they sought to remove what they regarded as a fundamental inconsistency, the 'thing in itself', from the critical philosophy. They achieved this by rejecting it, and transforming Kant's system into a consistent Idealism, which enabled one to deduce the basic types of matter and their laws of action.

Thorough-going and consistent Idealism regarded natural phenomena as the unfolding and self-revelation of the Absolute's creative reason, and accordingly required an animist universe, in which 'a general organism is the prerequisite of everything mechanical'.[40] This universal vitality coincided with a pullulating cosmic duality.

Some scientists were ambivalent in their relations with the high priests of *Naturphilosophie*, admiring their stress upon the philosophical foundations of natural science while deploring their indifference to scientific empiricism.[41] Yet even when they were unhappy with the animism of the true *Naturphilosophie*, all members of the 'idealist' school of chemistry[42] (which embraced both Kantians and post-Kantians) were in fundamental accord with Schelling that 'The first principle of a philosophical system is to *go in search of polarity and dualism throughout all of nature.*'[43] Overriding and encompassing this dualism was the fundamental idea of the unity of nature and of spirit, the reconciliation of seeming opposites in a higher Oneness.[44]

[37] ibid., vol. 1, p. 557: letter of 1 Jan. 1800.

[38] See Treneer, op. cit. (18), ch. 5.

[39] G. Eriksson (*Lychnos*, 1–37 (1965)) makes a clear distinction between Kantian and Romantic (more correctly, post-Kantian Idealistic) dynamism.

[40] F. W. J. von Schelling: cited by A. G. von Aesch, *Natural Science in German Romanticism*, p. 149. New York (1941).

[41] e.g. H. C. Oersted; see Chapter 4 below.

[42] Some aspects of this school of dynamical chemistry are discussed by D. M. Knight, Steps towards a dynamical chemistry. *Ambix* 14, 179–97 (1967).

[43] Quoted by R. C. Stauffer, *Isis* 48, 36 (1957).   [44] ibid., p. 39.

The application of these concepts to produce a theory of matter and of chemical affinity seemed to them a simple step to take:

Matter is the product of *forces* . . . Matter in general, or that which fills space, arises from the cooperation of three forces: the forward-striving, which appears as repulsion (first dimension); the retreating force, attraction (second dimension); and the synthesis of both these, manifesting itself as gravity (third dimension). Light raises these forces to a higher potential, where they act as the causes of *dynamic* processes, or of the specific differences of matter.[45]

The qualities of matter depend therefore upon the quantitative relations of the primary forces constituting the matter. Chemical affinity is another relation between heterogeneous bodies,

those namely, in one of which the relation between the primary forces is the reverse of that in the other. The resulting compound is the mean dynamical ratio of the primary forces which have been brought into activity during the process, and consequently its properties differ essentially from those of its elements.[46]

This goes beyond Kantian dynamism in its pretensions to quantification,[47] and far outdistances modest Newtonian dynamism. The chemistry of *Naturphilosophie* gained adherents, especially in Germany, but also brought upon itself intense hostility. Where Kantian or Newtonian dynamism was accepted, it was clearly differentiated from the outrageous speculations of Schelling and his followers. 'Idealistic' chemists fell into two groups. Swedish universities at the beginning of the nineteenth century clearly distinguished between dynamists of Kant's school and the new dynamists. The former were widely respected; the latter encountered very strong opposition.[48] It is necessary that the historian recognizes this distinction with equal precision.[49, 50] Davy himself seems to have failed to do so, thereby suggesting that he had no intimate acquaintance with any of the German Idealistic systems.

There are fragments suggesting that a philosophy of Idealism was stirring in Davy; in a notebook of about 1800, for example, he summed up a series of relationships in the phrase 'existence is unity'.[51] Such

[45] Quoted from Schelling by G. Hennemann, *Naturphilosophie im 19 Jahrhundert*, p. 35. Freiburg and Munich (1959). Electricity was similar in its effects to light; this was to prove of great significance (see Chapter 4 below).

[46] L. Gmelin, *Hand-Book of Chemistry*, trans. H. Watts, vol. 1, p. 159. 19 vols., London (1848).

[47] Kant never claimed that his (qualitative) suggestions about the causes of chemical properties were anything more than speculations: and he never attempted to formulate quantitative laws of chemical affinity.   [48] Eriksson, op. cit. (32).

[49] Davy's rejection of fluids has been connected with *Naturphilosophie;* see L. P. Williams, *Michael Faraday*, p. 68. Chapman & Hall, London (1965). Davy's ideas were, however, different in, e.g., *Phil. Trans. R. Soc.* **116**, 383–422 (1826); J. Davy, op. cit. (3), vol. 1, pp. 314, 318.   [50] See Chapter 1 and pp. 31–2 above.

[51] R. Instn MS. 15j.

remarks, however, can scarcely be used to identify Davy's philosophical allegiance—Kant, Schelling, and Cudworth are each equally likely as sources, and there is no reason to suppose that Davy would have owed special allegiance to any one of them.

Davy's slight familiarity with the German philosophy appears, for example, in his attack on Ritter (made around 1808), who had dared to propose theories and claim priorities in electrochemical science:

Ritter's errors as a theorist seem to be derived merely from his indulgence in the peculiar literary taste of his country, where the metaphysical dogmas of Kant which as far as I can learn are pseudo platonism are preferred before the doctrines of Locke and Hartley, excellence and knowledge being rather sought for in the infant than in the adult state of the mind.[52]

This seems clear testimony; it strongly suggests that, for Davy, German metaphysics was at best irrelevant to the pursuit of science. The phrase 'as far as I can learn' implies, moreover, that he had never personally studied Kant. The implication is reinforced by Davy's failure to indicate that Ritter was already under the influence of Schelling rather than of Kant.

But Davy, attacking Ritter, was simultaneously reacting against Coleridge.[53] One must remember also Davy's determination 'to mould himself upon the age in order to make the age mould itself upon him'.[54]

[52] R. Instn MS. (n.d., but about 1808). In the same lecture, Davy refers to 'Plato—who hiding philosophy in a veil of metaphysical tinsel fitted only to pamper the senses lived luxuriously with Parasites in the court of Dionysius the tyrant', in contrast with 'Anaxagoras in solitary exile . . . for having ventured to instruct the *Athenians* at that time unprepared for knowledge, in the truths of Nature'. So much for Platonism in science.

[53] The uneven relations between Davy and Coleridge can be traced through the latter's published letters (ed. Griggs) and notes (ed. K. Coburn). Davy's possible relations with Ritter's ideas remain obscure to me. Some passing references to Ritter's experimental results exist in Davy's notebooks (R. Instn MS. 13j, p. 106; MS. 20c, Clifton 1800, p. 82; MS. 22b, pp. 9, 19, 21, 42, 70), but their importance is not obvious. There is no evidence that Davy read German, he never made any acknowledgement to Ritter, and in any case Ritter's ideas were little known in England. (cf. J. Bostock, *An Account of the History and Present State of Galvanism*, p. 57. London (1818).) However, most of Ritter's principal chemical notions were at least reported in French or English journals. Davy's vanity would not readily own a debt to a somewhat obscure, albeit ingenious, foreigner, and there are many apparent coincidences between Davy's and Ritter's ideas. Both tended towards a dynamical theory of chemistry, both placed great emphasis on the role of electricity in chemical reaction, both believed in the fundamental simplicity of matter, and both flirted for a time with a modified phlogiston theory. Also, both considered themselves the founders and leaders of electrochemical science. I am almost sure that Davy did steal from his despised rival, but cannot prove my suspicions. In any event, Davy's fundamental ideas could have been developed in the same form had Ritter never existed. No works by Ritter found their way into the library of the Royal Institution while Davy worked there.

[54] K. Coburn, ed. *The Notebooks of Samuel Taylor Coleridge*, vol. 2, entry 1855, [Jan. 1804]. 2 + 2 vols., New York (1957–61).

At the beginning of the nineteenth century few Englishmen had studied Kant, and the philosophical climate in Britain was scarcely sympathetic towards Idealism. It is unlikely that Davy courted unpopularity by publicly praising a philosophy that he knew repelled most of his countrymen.[55] At various stages throughout his life, and notably in *Consolations in Travel*[56] (written well after his active scientific career had ended), Davy evinced marked Neo-Platonic tendencies, but these may have stemmed at least equally well from the British as from the German tradition.

Thus none of the arguments leads to an unequivocal link with a particular form of philosophic idealism. But Davy was unquestionably a Christian and a Newtonian; these facts, I believe, give us the keys to Davy's speculative thought, while any other approach merely leaves us with a puzzle. Even before Davy met Coleridge, he had made notes on natural theology and Newtonian physics; he read widely among British authors of the seventeenth and eighteenth centuries, and almost everywhere found support for his positions in religion and physics. Then, very possibly, he read Nitsch, and found there nothing to upset his basic assumptions. This is scarcely surprising, since Kant's philosophy was framed in a Newtonian cosmos, and in sympathy with Christian faith. Davy never *studied* Kant, but he possessed a good second-hand knowledge of his ideas, and saw in it a corroboration of precisely those elements that his own British background had provided already. There is no contradiction in Davy's apparent oscillation between English and German thought. He was steeped in the former—he was a patriot in matters of the intellect—and he selected complementary and matching elements from the latter. That is all.

If this conjectural interpretation is accepted, the metaphysical foundations of Davy's science are seen to be all of one piece. His dynamism is Newtonianism of the eighteenth century and it ultimately controls his theories of affinity and of the chemical elements.

Electrical theories of affinity

Let us turn now to another phase of Davy's speculative thought. In the early years of their friendship, Coleridge wrote to Davy asking for his metaphysical ideas on thought and sensation[57]—and Davy's notes were here decidedly in the British, not the German, metaphysical tradition. 'Quere? does not the blood in circulating through the brain

[55] See, for example, *Edinb. Rev.*, **1**, 254–5 (1803); F. Jeffrey, *Edinb. Rev.* **17**, 167 (1810).

[56] H. Davy, *Consolations in Travel; or, The Last Days of a Philosopher*. London (1830); see pp. 60–2 below.

[57] Griggs, op. cit. (22), vol. 1, p. 590: letter of 7 June 1800.

and nerves supply the ultimate atoms to the mind which constitutes ideas . . .'[58] Davy's ideas were at this time compounded of Hartley's materialism, Newtonianism (its theories of matter and of natural theology), dissatisfaction with Lavoisier's *Elements*, and the boundless ambition of youth. 'Yet are my limbs with inward transports thrilled, and clad with new-born mightiness around.'[59] This amalgam was, in 1800, the essential Davy. His restless activity would not be satisfied merely with criticizing; he must be doing. But—doing what? Dissatisfaction with the French system of chemistry meant tackling it at its weakest points, splitting and toppling its table of elements[60]—an enterprise wherein Beddoes encouraged him.[61] But how was this to be accomplished? In 1800, a new tool for overcoming the affinities of bodies burst upon the scientific world, with Volta's announcement of his discovery of the electric pile.[62] It was to provide Davy with his greatest successes.

Volta's invention of the battery at once opened new vistas for chemical research. Here at last was the tool for investigating the elements and internal structure of bodies. Joseph Priestley had predicted this role for static electricity;[63] Humphry Davy, his appointed heir, was to fulfil his prediction with current electricity.

Sir Joseph Banks, President of the Royal Society of London, showed Carlisle Volta's letter announcing his discovery. With Nicholson, Carlisle at once constructed a pile, and on 2 May 1800 used it to decompose water. They were astonished to find that hydrogen and oxygen were evolved at different electrodes. 'This new fact still remains to be explained, and seems to point to some general law of the agency of electricity in chemical operations.'[64] In the same year, Haldane,[65] Cruickshank,[66] and Henry[67] were among those who followed up this initial and perplexing experiment. Davy at once began his own researches, delighted with the new tool that Volta and the others had put

[58] R. Instn MS. 13c (*ca.* 1799–1800).   [59] ibid.   [60] R. Instn MS. 20b.

[61] One should not neglect the possible influence of Dr. Beddoes on Davy's ideas and researches concerning the chemical elements. Beddoes's 'Specimen of an Arrangement of Bodies according to their Principles' in his *Contributions to Physical and Medical Knowledge, principally from the West of England* (Bristol, 1799) rejects caloric, talks about phosoxygen, and suggests that the alkalis and alkaline earths are compounds containing oxygen or phosoxygen. However, on page 212 of the work, Beddoes asserts that Davy's rejection of caloric, ideas about phosoxygen, etc., were original; so perhaps Beddoes merely encouraged a pupil who had independently come to think along the same lines as himself. Beddoes seems to have had many irons in many fires—I suspect that he had more influence on young Humphry Davy than has hitherto been demonstrated.

[62] A. Volta, *Phil. Trans. R. Soc.* **90**, 403 ff. (1800).

[63] J. B. Priestley, *The History and Present State of Electricity*, preface. 2nd edn, London (1769).

[64] *Nicholson's J.* **4**, 179–87 (1800).   [65] ibid., **4**, 242–5 (1800).
[66] ibid., **4**, 187–91 (1800).   [67] ibid., **4**, 223–6, 245–9 (1800).

into his hands—it was as if they had done their work specially for him![68]

First at Clifton, and then at the Royal Institution, Davy eagerly pursued his galvanic investigations. While still at Clifton he was able to report that 'I have met with unexpected and unhoped-for success. Some of the new facts on this subject promise to afford instruments capable of destroying the mysterious veil which Nature has thrown over the operation and properties of ethereal fluids.'[69] It was not long before he concluded, in contradiction to Volta's theory,[70] that the electrical effects in the piles were due to the contact of dissimilar metals, and that chemical changes were responsible for electrical effect.[71]

By September 1801 he had refined this statement, and started to relate chemical affinity to electrical power; at the same time he had begun his synthesis of the chemical and contact theories of voltaic action. An instantaneous circulation of 'galvanic influence' might be produced by contact, but its permanent excitation required 'a certain exertion of . . . chemical affinities'. This was the conclusion from an induction based on observations of a number of different cells of the same construction but containing different electrodes and electrolytes. Davy had noticed that the most powerful circuits were those where the electrodes differed most widely in their oxidability, and that there was a direct relation between the galvanic power produced, and the 'intensity of the primary chemical agencies' exerted. He interpreted the latter in terms of differing affinities of the electrodes for oxygen, and suggested that these were correlated with electrochemical properties. Finally he implied that, within the battery, there was a common cause for chemical and electrical changes.[72]

Meanwhile, Wollaston[73] had repeated the decomposition of water by electric sparks,[74] as Deiman and Troostwyk,[75] Pearson,[76] and others had done before him; in 1805, Davy heard of Biot's experiments on the supposed synthesis of water by the compression of a mixture of hydrogen and oxygen. Finally, there was the standard experiment of the

[68] cf. Siegfried, Royal Institution discourse, 3 May 1968.
[69] H. Davy to D. Giddy, 20 Oct. 1800: quoted by Paris, op. cit. (25), vol. 1, pp. 109–10.
[70] Volta, op. cit. (62).
[71] Nicholson's J. **4**, 337–42 (1800).
[72] H. Davy, op. cit. (17), vol. 2, pp. 188–209.
[73] Phil. Trans. R. Soc. **91**, 427–34 (1801). Davy (see below) rightly saw that this was not a case of Voltaic decomposition. Subsequent authors did not always make this distinction, e.g. Faraday, Experimental Researches in Electricity, vol. 1, para. 309. 3 vols. London (1839, 1844, 1855).
[74] Obs. Phys. 369 ff. (1789). Volta suggested and described similar experiments in his correspondence with M. van Marum, Secretary of the Hollandsche Maatschappij der Wetenschappen (Van Marum archives, Volta to Van Marum, 28 Nov. 1782, 23 July 1789, 3 March 1802).
[75] Nicholson's J. **1**, 241 ff. (1797).   [76] Phil. Mag. **21**, 362–4 (1805).

composition of water from its elements by sparking their mixture. All three experiments, to Davy, exemplified *mechanical* means of chemical synthesis and decomposition: mechanical shock could approximate or separate the particles of matter, thereby effecting changes in their chemical state. For example, in Biot's experiment, 'the particles being brought nearer to each other ceased to be mutually repulsive and were brought within the limits of elective affinity'.[77] This clearly presupposes at least a partly dynamical theory of matter, perhaps something similar to Rowning's.[78] On these terms, affinity was not itself a primary force, but rather a mere consequence of the attractive powers of matter, which could be exerted when the repulsive ones had been overcome.

The two preceding paragraphs contain both an incipient electrical theory of chemical affinity, and a purely mechanical one. Davy had to reconcile them with one another, either by modifying the generally accepted notion of mechanical forces, or by acquiring an equally unconventional concept of electricity. In either event, the reconciliation could only occur within a dynamical framework.

Davy continued to busy himself with experiments on galvanism, but it was not until the autumn of 1806 that his 'power superior even to will'[79] clearly emerged for the first time, in the Bakerian Lecture 'On some chemical Agencies of Electricity',[80] a paper that is absolutely central to any consideration of affinity theory in the first half of the nineteenth century. Its brilliance and significance were immediately recognized—Berzelius called it 'one of the best memoirs which has ever enriched the theory of chemistry',[81] while the *Edinburgh Review* acclaimed the investigation as unsurpassed 'in modern times, for closeness, copiousness, and minute accuracy'.[82] Coleridge was told by 'a fellow of the *Royal Society*' that the sensation produced by the reading of Davy's paper there 'was more like stupor than admiration—and the more as the whole train of these discoveries have been the result of profound Reasoning, and in no wise of lucky accident'.[83]

[77] R. Instn MS. 1.   [78] See p. 10 above.

[79] 'Restlessness of thought, power superior even to will, ardent, but indefinite hope— these constitute the great elements of that feeling which always has something above the common habits of thought; has been, as it were, supernaturally infused into the mind, or self-born in it . . .' H. Davy, notebook entry of Feb. 1805: cited by J. Davy, *Fragmentary Remains Literary and Scientific of Sir H. Davy*, p. 115 London (1858).

[80] *Phil. Trans. R. Soc.* **97**, 1–56 (1807).

[81] J. J. Berzelius, *Traité de chimie*, trans. Jourdain and Esslinger, vol. 1, p. 164. 1st Swedish edn (*Lärbok i kemien*), Stockholm (1818), this edn, 8 vols., Paris (1829–33).

[82] H. Brougham (*Edinb. Rev.* **11**, 398 (1808)) added: 'It is no small proof of Mr. Davy's natural talents and strength of mind, that they have escaped unimpaired from the enervating influence of the Royal Institution; and indeed grown prodigiously in that thick medium of fashionable philosophy.'

[83] Coleridge to Dorothy Wordsworth, 24 Nov. 1807: cited by Treneer, op. cit. (18), p. 104.

In order to bring out certain assumptions underlying Davy's work, it will be necessary to dwell on the theoretical content of this lecture. Davy began by proving, in a series of brilliant and meticulous experiments, that the electrolysis of water did not generate any new matter, and that the only products evolved were oxygen and hydrogen.[84] (This was necessary to refute the claims of those who, like Ritter,[85] Gren,[86] and Brugnatelli,[87] considered that electricity was the material, and not merely the instrumental cause of electrolysis.) He went on to examine numerous instances of the electrolysis of saline solutions, and found that in every case the constituent parts of the bodies were 'newly arranged by the effects of electricity'.[88] It also emerged that some substances were consistently attracted to the anode and repelled by the cathode, while others were affected in precisely the opposite manner. Rearrangement and transfer arose because 'these attractive and repulsive forces are sufficiently energetic to destroy or suspend the usual operation of elective affinity'. These energies, Davy suggested, were communicated from particle to particle, 'so as to establish a conducting chain in the fluid' terminating at the poles. Attraction and repulsion from the poles, transmitted by means of this conducting chain, produced locomotion, and brought about a succession of decompositions and recompositions, which enabled one to account for the phenomena both of transfer and of electrolytic decomposition.[89]

The next stage in the investigation was the determination of the electrical states of insulated bodies, before and after their contact with metals. Davy found, as he had expected, that dry acids received a negative charge after contact with metals, while dry alkalis received a positive charge; 'the effect was exceedingly distinct'.[90] From these results, Davy concluded that acids, alkalis, and metals all existed in determinate electrical relations with one another. He immediately grasped at the analogy between chemical and electrical attraction:

In the present state of our knowledge, it would be useless to speculate on the remote cause of the electrical energy, or the reason why different bodies,

[84] *Phil. Trans. R. Soc.* **97**, 12 (1807).

[85] *Chem. Ann.* **1**, 41–63 (1801).

[86] *Hollandsche Maatschappij der Wetenschappen*, Van Marum archives, Gren to Van Marum, 12 June 1791.

[87] *Ann. Phys.* **8**, 284 (1803).

[88] *Phil. Trans. R. Soc.* **97**, 15 (1807).

[89] ibid., 29; cf. the theory propounded by F. T. von Grotthus, *Ann. Chim. Phys.* **58**, 55 (1806). This was where Faraday and Davy parted company—Faraday was to show it was unnecessary to invoke poles, sources of attraction and repulsion, to account for the phenomena of electrolysis. Instead, one had to consider the arrangement of molecular forces throughout the conducting medium. See below.

[90] *Phil. Trans. R. Soc.* **97**, 35 (1807).

after being brought into contact, should be found differently electrified; its relations to chemical affinity is, however, sufficiently evident. May it not be identical with it, and an essential property of matter?[91]

He suggested that different particles in combination must be supposed to preserve 'their peculiar states of energy'. This would mean that compounds of substances of unequal opposite charges would remain electrically charged—an inference that seemed to be borne out by experiments demonstrating the existence of charged compounds.

Within the context of the electrical states of matter, Davy advanced towards a theory of elective affinity that would account not merely for the general phenomenon of attraction, but also for its specificity.

When two bodies repellent of each other act upon the same body with different degrees of the same electrical attracting energy, the combination would be determined by degree; and the substance possessing the weakest energy would be repelled; and this principle would afford an expression of the causes of elective affinity, and the decomposition produced in consequence . . .[92]

In more complicated cases than the three-body problem where

$$A + BC \rightarrow AB + C,$$

there was a quantitative balance of the attractive and repellent powers. The notion of an equilibrium of the attractive and repulsive inherent powers of matter, leading to a stable but complex configuration of bodies, was compatible with many theories of matter, including several variants of Newtonianism. In particular, it was compatible with the notion of universal attraction that Berthollet thought was the basis of chemical as well as celestial attraction. Davy therefore found it easy to reconcile his electrical theory of affinity with Berthollet's concept of mass action;[93] for, he tells us,

the combined effects of many particles possessing a feeble electrical energy, may be conceived equal or even superior to the effect of a few particles possessing a strong electrical energy . . .

Allowing combination to depend upon the balance of the natural electrical energies of bodies, it is easy to conceive that a measure may be found of the artificial energies, as to intensity and quantity produced in the common electrical machine, or the VOLTAIC apparatus, capable of destroying this equilibrium; and such a measure would enable us to make a scale of electrical powers corresponding to degrees of affinity . . . [Finally, the effect of] HEAT, in producing combination, may be easily explained . . . It not only gives more freedom of motion to the particles, but in a number of cases it seems to exalt the electrical energies of bodies . . .[94]

[91] ibid., 39.   [92] ibid., 41.
[93] See Chapter 7 below.   [94] op. cit. (88), 43.

If the worth or usefulness of a scientific explanation is measured by the number of previously unrelated phenomena that it brings together in a unified, cross-linked pattern,[95] then this paper must be accounted among the most valuable ever. In a few pages and with an astonishing sureness of touch, Davy had constructed a theory connecting the disparate phenomena of static and galvanic electricity, electrolysis,[96] the forces of chemical affinity and chemical combination, the heat and light of chemical reaction, affinity tables, and the electrochemical series, and had shown the possibility of chemists arriving, assisted by the boundless power of Volta's battery, at a knowledge of the true elements of bodies. The preceding pages have concentrated on those aspects of the paper that are germane to affinity theory; but its significance was far wider. Davy himself recognized this quite distinctly.

The Voltaic Battery was an alarm bell to the slumbering energies of experimenters in every part of Europe, and it served no less for demonstrating new properties of Electricity and for establishing the laws of this Science, than as an instrument of discovery in other branches of knowledge; *exhibiting relations between subjects before apparently without connection and serving as a bond of unity between chemical and physical philosophy.*[97]

Davy has dropped the notion of electrical fluids; electricity is presented as a force essential to matter, while matter itself is particulate. The whole picture is well-guarded by vagueness,[98] and all that one is justified in saying is that Davy *may* be thinking in purely dynamical terms, but that a literal interpretation would suggest the perennial compromise of dynamical atomism (the phrase is Oersted's).[99]

In the following year, Davy, volatile as ever, told a lecture audience at the Royal Institution that electricity was a substance, though perhaps rather a refined creature of the intellect than a matter of 'common

[95] A. Arber, *The Mind and the Eye*, p. 59. Cambridge (1964).

[96] For a good account of Davy's theory of electrolysis, see C. A. Russell, *Ann. Sci.* **15**, 1 (1959).

[97] R. Instn MS. 1 (n.d.). (The italics are mine.) Whewell also was to regard galvanism as the bridge between physics and chemistry. (*History of the Inductive Sciences*, vol. 3, pp. 98 ff.) 3 vols., London (1837).

[98] cf. Faraday's comment (op. cit. (73), vol. 1, section 482): 'The facts are of the utmost value, and, with the general points established, are universally known.' (Faraday was writing in 1833.) '*The mode of action* by which the effects take place is stated very generally, so generally, indeed, that probably a dozen precise schemes of electro-chemical action might be drawn up, differing essentially from each other, yet all agreeing with the statement there given.' As Sir Harold Hartley observes (op. cit. (4), p. 54), Davy was wise in avoiding further speculation.

[99] Oersted propounded his dynamical atomism in his correspondence with Weiss and Ritter (*Correspondance de H. C. Oersted*, ed. Harding, vol. 1, pp. 282–326. 2 vols., Copenhagen (1926)), and in his *Ansichten der chemischen Naturgesetze*, pp. 252 ff. Berlin (1812).

sensation'.[100] Much as his ideas on the imponderables may have vacillated, he never wavered in his conviction that nature was simple, and that the chemical elements were therefore few. In January 1807 he introduced a lecture on the 'chemistry of nature' with the statement that 'all natural bodies consist of *different arrangements or combinations of a few simple parts or elements*, and it is on a knowledge of the *invariable properties and agencies* with which these elements are endowed, that the whole of the demonstrative part of the science depends'.[101]

He was determined to find these few simple elements. The electrical decomposition of the fixed alkalis, and the discovery of sodium and potassium[102] ('Cap¹· Expt·'), followed swiftly on his first Bakerian Lecture. Davy predicted the analogous decomposition of the alkaline earths,[103] and extended the argument to alumina and silica:

. . . there is no small reason to hope, that even these refractory bodies will yield their elements to the methods of analysis by electrical attraction and repulsion.

In the electrical circuit we have a regular series of powers, of decomposition from an intensity of attraction, so feeble as scarcely to destroy the weakest affinity existing between the parts of a saline neutral compound, to one sufficiently energetic to separate elements in the strongest degree of union, in bodies decomposable under other circumstances.[104]

As the electrical power available to mankind increased, so increasingly refractory bodies could be decomposed—once again, the notion of an electrochemical series, related to the order of affinity, was implied.[105] At this stage Davy still thought that oxygen was the only truly negative element, so that bodies attracted to the positive pole must either be oxygen, or contain it in electrical excess; bodies attracted by negative electricity were pure combustibles, or contained an excess of combustible matter. Although he had breached the wall of Lavoisier's table of elements, he was not yet ready for his major assault. The muriatic acid radical (chloride ion) was negative in electrolysis, so it probably contained oxygen.[106]

[100] H. Davy, *Collected Works*, ed. J. Davy, vol. 8, p. 239. 9 vols., London (1839).

[101] ibid., vol. 8, p. 168. (The italics are mine.)

[102] *Phil. Trans. R. Soc.* **98**, 1–44 (1808).

[103] H. Davy, in a MS. lecture at the R. Instn, op. cit. (100), vol. 5, p. 103 n.) described the analogical train of thought, beginning from the compound nature of ammonia, which led him to attempt the decomposition of the alkalis. He was not entirely alone in early suspicions of their compound nature. See for example, *Nicholson's J.* 513 (1801): 'M. Werner has always thought that the alkalis and the earths were compounds, founded on the incontestable facts furnished by geognostic observations; there can be no doubt that silica, alumina, etc., are not equally compounds.'

[104] H. Davy, op. cit. (102), p. 42.

[105] cf. Berzelius, Chapter 5 below.

[106] H. Davy, op. cit. (102), p. 43

The voltaic pile, clearly, could overcome the affinities of bodies. Davy saw with equal clarity that the 'powers and affinities of the new metals' (sodium and potassium) also could serve as powerful analytical tools, not least because of the alkali metals' great affinity for oxygen.[107] His ideas about the chemical elements and theories of chemical affinity were intimately and inextricably combined.

At this juncture, Davy was seized with a dangerous illness. In March 1808 he resumed his lectures; though he had been unable to continue with laboratory work in the interval, his mind had been as restlessly active as ever.[108] He returned to the Royal Institution with his old conviction of the power and potential of analogical thought in making discoveries—'analogy is in science what blossom is in vegetation, beautiful and replete with promise, and may ripen into useful fruit'.[109] But analogy, even direct correlation, was no proof of the identity of chemical and physical forces. Change underlay all natural operations, change implied motion, and motion was, for Davy, evidence of the existence of active powers.

In our discussions on every part of the Science of Nature we are obliged to recur to these powers—But though sound philosophy permits us to allow them as *active* it is far from considering them inherent in or necessarily attached to matter.—After the example of Newton they may be considered as principal but not ultimate, as secondary but not primary causes . . .
Every species of attraction and repulsion that we are acquainted with *may* be the result of one grand and universally operating Law; and the farther we advance in Science the more this idea becomes probable. At present however it is necessary to make distinctions between gravitation, cohesion, magnetic and electrical attraction, and to follow the facts awaiting patiently the progress of discovery and the perfection of analogies.[110]

Davy, imbued with this mature caution, became less confident of the identity of chemical with electrical attraction. It was possible but by no means certain that they were identical; Davy could, however, be sure that 'the facts of the dependence of the chemical arrangements of matter upon electrical functions will be permanent', and that there was at least a positive correlation between the electrical states of bodies and their roles in combustion.[111] This caution was maintained in his paper on the decomposition of the earths,[112] where he contented himself with indicating the coincidence of certain electrical and chemical states of matter:

Thus acids are uniformly negative, alkalies positive, and inflammable substances highly positive: and as I have found, acid matters when negatively

[107] H. Davy, op. cit. (102), p. 44.   [108] H. Davy, op. cit. (100), vol. 1, p. 114.
[109] ibid., vol. 8, p. 317.   [110] R. Instn MS. 2: lecture of 19 March 1808.
[111] H. Davy, op. cit. (100), 8, p. 284.   [112] *Phil. Trans. R. Soc.* **98**, 333–70 (1808).

electrified, seemed to lose all their peculiar properties and powers of combination. In these instances, the chemical qualities are shewn to depend upon the electrical powers; and it is not impossible that matter of the same kind, possessed of different electrical powers, may exhibit different chemical forms.

I venture to hint at these notions; but I do not attach much importance to them; the age of chemistry is not yet sufficiently mature for such discussions; the more subtile powers of matter are but just beginning to be considered; and all general views concerning them, must as yet rest upon feeble and imperfect foundations.[113]

Davy was hinting strongly here at a dynamical theory of matter. However, he resolutely refused to commit himself to such a theory, or to the notion that chemical and electrical powers were one and the same. Even though such a theory 'perfectly coincides with the phenomena, and was not a suggestion of the mind *a priori*; but flowed irresistibly from the facts yet still I do not wish it to be considered in any other light than as a probable hypothesis'.[114]

Davy believed that matter was ultimately simple—there might be, he suggested, one or two species of ethereal matter and two or three species of ponderable matter;[115] these would combine to form bodies whose electrical energies were susceptible of measurement, and which corresponded to their chemical attractions and repulsions—'so the laws of affinity may be subjected to the forms of the mathematical Sciences and the possible results of new arrangements of Matter become the objects of calculation'.[116]

The next year, 1809, saw a continuation of Davy's electrochemical researches, published as his fourth Bakerian Lecture, and saw also an increasing flood of speculation about the nature of matter, natural forces (including chemical affinity), the imponderables, and the chemical elements—a torrent of inextricably interwoven hypotheses, relinquished as readily as assumed.

The only use of an hypothesis is, that it should lead to experiments; that it should be a guide to facts. In this application, conjectures are always of use.

[113] ibid., cf. T. Young, *A Course of Lectures on Natural Philosophy and the mechanical arts*, vol. 1, p. 684. 2 vols., London (1807):
. . . perhaps . . . time may show, that electricity is very materially concerned in the essential properties which distinguish the different kinds of natural bodies . . .; but at present it is scarcely safe to hazard a conjecture on a subject so obscure, although Mr. Davy's experiments have already in some measure justified the boldness of the suggestions.
[114] R. Instn MS. 2.
[115] J. Davy, *Memoirs of the Life of Sir Humphry Davy*, vol. 1, pp. 403 ff. 2 vols., London (1836).
[116] R. Instn MS. 2: lecture 10 of course on electrochemical science.

The destruction of an error hardly ever takes place without the discovery of truth . . .

Hypothesis should be considered merely an intellectual instrument of discovery, which at any time may be relinquished for a better instrument.[117]

For all their expendability, Davy's hypotheses were essential to the growth of his concepts.

Whereas Davy had previously toyed with ideas about imponderable fluids—one of the major reasons for his early enthusiasm about the voltaic pile had been its potential for elucidating 'the philosophy of the imponderable, or ethereal fluids'[118]—in this year he veered strongly away from them. The intense ignition of a platinum wire in a vacuum convinced him that there was no specific fluid of heat—where did the heat come from in this experiment? Not from the vacuum, nor, certainly, from the metal. On the supposition that it arose from the combination of two electrical fluids, there ought to be a corresponding production of cold in some part of the system; but this did not happen. 'I once had this idea. It satisfies the imagination, but not the reason.' And so he disposed of the alternative fluid theories, one by one, until he burst out against the whole tribe of imponderable fluids: 'Vulgar idea—like that of a peasant, everything done by a spring; so every thing must be done by a fluid. The ether was the ancient fluid; then there was a phlogistic fluid: we have had the magnetic fluid, the vitreous fluid, the resinous fluid', and so on—they are all nonsense![119]

This firm rejection of imponderable fluids in general, and of electrical fluids in particular, supported his dynamical or force theory of chemical affinity. Whatever might be the ultimate truth, a clear simplification was achieved by the identification of chemical with electrical attraction; and 'our arrangements are most likely to be *true*, when they are most characterized by simplicity'.[120] Now, for the first time, Davy recognized publicly the implications of his theory of affinity for theories of electricity—fluid theories of electricity were incompatible with the force theory Davy propounded, concerned as it was with states of matter, exaltation of attractive energies, and mutability of chemical properties under electrical influences.[121] Quite apart from these grounds, which might reasonably be attacked as stemming merely from convenience (fluid theories of electricity would not suit Davy's notions on matter and chemical affinity, so there could be no electrical fluid), Davy considered that he had one powerful and decisive argument against fluids—they were superfluous, they violated his criterion of simplicity, and so they must be eliminated.

[117] H. Davy, op. cit. (100), vol. 8, pp. 346–7.
[119] H. Davy, op. cit. (100), vol. 8, p. 348.
[121] ibid.

[118] *J. R. Instn* **1**, 50 (1802).
[120] R. Instn MS. Dec. 1809.

The simplicity criterion was taken from Newton,[122] as was the complementary opinion that all matter was ultimately and essentially the same.[123] Davy, as we have seen, sought to reconcile this underlying unity with the sensible diversity of matter by suggesting that the prime matter in different electrical states might appear as different chemical substances; arrangements and different attractions of the parts of matter would yield varied forms.

... Should Chemistry hereafter confirm this idea, it would become the most *important*, and *noblest* of the Sciences;—for it would refer the diversified and multifarious phaenomena of ye terrestrial universe, to powers as simple and uniform, as those which govern the movements of the heavenly bodies;—And thus the *History* of Nature, would form *one System*, as Nature itself forms one intelligent design; and all the changes of matter might be referred to *one law* . . .[124]

Thus natural theology, with Davy as with Newton, was the essential justification of the use of Ockham's razor. It is necessary to insist upon the relation of Davy's natural philosophy to his natural theology; his theology came to appear to him as a major source and justification for his philosophy. It is worth remarking that such an attitude was not typical among nineteenth-century physical scientists, and that in Davy and Faraday we see the culmination of the mutual dependence of theories of matter and of natural theology, the true end of the Newtonian tradition.[125] (There were also those who, like Laplace,[126] attempted a complete separation of the two lines of thought; their mechanics was Newtonian but their philosophy was not.)[127]

To return to Davy and chemical affinity: Davy also adopted from Newton's *Opticks* the conviction that chemical affinity was strongest between the simplest bodies, and that as bodies increased in complexity, so their affinities became weaker.[128] Nitrogen was notoriously inert. Davy believed that elements possessed active powers, so that they could not be inert. Thus it seemed reasonable to assume that nitrogen was compound;[129] the amalgamation of ammonia, which

[122] e.g. J. Newton, *Opticks*, p. 397. 4th edn, London (1730), repr. Dover, New York (1952): 'And thus Nature will be very comformable to herself, and very simple . . .'
[123] In connection with this aspect of Newton's theory of matter, see the article by J. E. McGuire, *Ambix* **14**, 69–95 (1967).
[124] R. Instn MS. 4: December 1809.
[125] cf. chapter 3, section 2.
[126] This attitude is exemplified by the (probably apocryphal) reply Laplace is said to have given to Napoleon's query about the place of God in celestial mechanics: 'Sire, I have no need of *that* hypothesis.'
[127] The eighteenth century attempted sporadically to purge metaphysics and theology from 'Newtonian' natural philosophy.
[128] Newton, op. cit. (122), p. 259.
[129] R. Instn MS. 3. However, Davy was careful to recognize probabilities for what

Davy had first investigated, and which Berzelius and Pontin had efficiently achieved,[130] was further evidence of the compound nature of this supposed element. Davy and Berzelius corresponded eagerly on this topic, vying with each other in experiment and error. The argument from affinity spurred on Davy in his efforts.

Quere may not Hydrogene and Nitrogene and Ammonia be all forms of the same species of Matter—combined or energetic in consequence of different electrical powers?[131]

I have come to [the] conclusion . . . that water is the basis of all the gases, and that oxygen, hydrogen, nitrogen, ammonia, nitrous acid etc., are merely electrical forms of water.[132]

From nitrogene I have obtained a large quantity of oxygene; but as yet I have not been able to conclude concerning its basis. It seems resolved into nothing but oxygene and hydrogene.[133]

Berzelius, misinterpreting the reaction of sodium with ammonia to form sodamide, and the subsequent hydrolysis leading to ammonia and caustic soda, went so far as to claim that ammonia contained more than $31\frac{1}{2}$ per cent of oxygen.[134]

Berzelius's mistakes arose from the application of his ideas on chemical combining proportions, Davy's from his subscription to Newton's '31st Query'. In his Bakerian Lecture for this year, 1809,[135] Davy was to attempt the union of two aspects of quantifiable chemistry—affinity and the law of equivalents, or Dalton's law of multiple proportions—in one comprehensive system. He used as premises for this synthesis two postulates that he had previously brought before the Royal Society. The first stated that chemical affinity and electrical attraction might be 'different exhibitions of the same property of matter'. The second stated that the attraction of acids for bases was inversely proportional to the quantities of oxygen they contained: this followed from the more

---

they were, and not to confuse them with certainties. In 1813 he wrote to Berzelius that the composition of nitrogen remained hypothetical. It was *very probable* that nitrogen contained oxygen; 'but it is absolutely necessary to distinguish between what is *very probable* and what is *known*'. See Söderbaum, ed. *Jac. Berzelius Bref*, vol. 2, p. 59. 6 vols. + 3 suppl., Uppsala (1912–32).

[130] *Ann Phys.* **26**, 198 (1810).

[131] R. Instn MS. 13j. n.d.

[132] H. Davy to T. A. Knight: cited by J. Davy, op. cit. (79), p. 129. cf. H. Davy, *Elements of Chemical Philosophy*, p. 486. London (1812). Such ideas are reminiscent of Ritter's.

[133] Letter from Davy to Berzelius, 20 March 1809: cited in Söderbaum, op. cit. (129), vol. 2, p. 10.

[134] Letter from Berzelius to Davy, 30 June 1809: quoted in Söderbaum, op. cit. (129), vol. 2, p. 13.

[135] H. Davy, op. cit. (100), vol. 5, pp. 225–82.

general postulate that oxygen and combustibles attracted one another according to the degree of their electronegativity and electropositiveness respectively.[136]

If one then supposed that the amount of base dissolved by an acid gave a measure of their mutual attraction, 'it would be easy to infer the quantities of oxygen and metallic matter from the quantities of acids and of basis in a neutral salt'. Davy claimed that he had used this argument in 1808 to predict that barytes would contain the least oxygen of all the alkaline earths; this had been confirmed by experiment. The same proportions, he added, would follow from an application of 'Mr. Dalton's ingenious supposition' of the existence of multiple combining proportions. Davy confessed in a footnote[137] that his principle of acid–base affinity being inversely proportional to oxygen content, 'though gained from the comparison of the electrical relations of the earths and their chemical affinities, in its numerical applications, must be considered merely as a consequence' of Dalton's law. In the spring of 1808 Dalton had sent Davy a list predicting the gravimetric composition of the alkalis and earths, and these predictions had accorded well with experiment. Davy's theory appears in the light of this admission as a *post facto* rationalization of Dalton's figures in terms of his own electrochemical theory. The significant point is that Davy recognized the need to integrate his theory and Dalton's—nature was 'very conformable to herself'.[138] Davy, like Kirwan before him, was confusing affinities with equivalents;[139] but unlike Kirwan he had a hypothesis to account for his experimental results.

In the same lecture, Davy, always aiming at a unified natural philosophy, also related chemical affinity to the mechanical properties of bodies; the condensation when bodies combined was greatest for those bodies that had the greatest affinities for one another; oxygen had a higher affinity for potassium than for sodium, so that the specific gravity of potassium oxide was necessarily greater than that of sodium oxide.[140, 141]

Davy's analytical researches in the next two years convinced him of the truth of the laws of definite and multiple proportions, although the basis of these laws was not in any 'speculations upon the ultimate particles of matter';[142] it lay, more soundly, in the mutual decompositions of neutral salts, and above all 'in the decompositions by the Voltaic

[136] cf. H. Davy, *Phil. Trans. R. Soc.* **98,** p. 41 (1808).
[137] H. Davy, op. cit. (100), vol. 5, p 272 *n.*
[138] cf. note (122) above.
[139] J. R. Partington, *A History of Chemistry*, vol. 3, p. 665. London (1962).
[140] H. Davy, op. cit. (100), vol. 5, p. 235.
[141] Some kind of dynamism underlies this argument.
[142] H. Davy, op. cit. (100) vol. 5, pp. 140 ff.
A M—E

apparatus, where oxygene and hydrogene, oxygene and inflammable bodies, acids and alkalies, etc., must separate in uniform ratios'.[143] When he wrote to Berzelius on this topic,[144] Davy expressed combining proportions in purely numerical terms, and made no mention of atomic hypotheses. When he did allude to Dalton's atomic notions, it was generally to oppose them, 'because I am convinced that they are contradicted by refined observation and opposed by the results of minute and accurate experiments'.[145] The fact of the existence of definite proportions militated against Berthollet's views on affinity;[146] naive Newtonian attraction, with which Berthollet broadly identified chemical affinity,[147] was incapable of accounting for the elective aspect of affinity, for it was an indiscriminate and general agent. Davy believed now that the acceptance of Berthollet's ideas on affinity would destroy all certainty in chemical science. If Berthollet were right, chemical combinations would be variable and complicated in composition, and all the material world would be irresistibly drawn to an equilibrium, connected with 'a quiescence and an eternal sleep in Nature';[148] a chaotic, disordered sleep: but 'the refined analogies of Science tend to confirm and not to confuse those ideas of order and design' that are so clearly evident in the world.[149] The impasse was reached because Davy and Berthollet had quite different concepts of the nature of a compound and of chemical combination: Davy would not have agreed with Berthollet that solutions were examples of chemical compounds,[150] and their different use of terms would have meant that they simply talked through one another. Berzelius criticized Davy for his rejection of mass action, and attempted to reconcile the concept of mass action with his electrical theory of affinity;[151] however, this reconciliation could be achieved only by removing mass action from the theoretical framework in which Berthollet had conceived it, and so Berzelius failed to dissuade Davy from his hostility towards it. Davy, in his first Bakerian Lecture, proposed an electrical theory of mass action. This was plausible, especially so because Davy did not try to specify the laws governing mass action. The particular patterns of force permitted and required by a given dynamical system might indeed lead to specific combinations in

[143] H. Davy, op. cit. (100), vol. 5, p. 328 n.

[144] Davy to Berzelius, 24 March 1811: quoted in Söderbaum, op. cit. (129), vol. 2, p. 22.

[145] R. Instn MS. 1.

[146] Hence Davy's aversion from mass action, which (see p. 39 above) he had formerly championed.

[147] C. L. Berthollet, *Essai de Statique Chimique*, vol. 1, p. 1. 2 vols., Paris (1803).

[148] R. Instn MS. 1.

[149] R. Instn MS. 4: 2nd lecture of course of Chemical Philosophy, Dublin, 1811.

[150] Berthollet, op. cit. (147), vol. 1, pp. 59–60.

[151] Letter of 1812 or 1813: quoted in Söderbaum, op. cit. (129), vol. 2, p. 41; Berzelius, op. cit. (81), vol. 4, p. 583–9.

definite proportions. A gravitational theory of affinity and mass action, however, could not be brought into alignment with the facts of chemical analysis.

Chemical attraction and gravitational attraction were not identical. Davy now amended a lecture to make a similar distinction between chemical and electrical attraction, for he felt that his theory had been 'often misunderstood and generally misrepresented'.[152] More emphatically than before, he insisted that the two forms of attraction were '*independent* phenomena; resulting from the same cause'. All the active powers of matter arose from the same cause,

which as it is intelligent must be divine . . . The *different powers in nature* as harmoniously cooperating must be referred to one source—When we perceive a great and magnificent Building we know that it must have been the result of the labours of various hands;—but we are secure that the perfect design by which its parts have been arranged must have been the combination of a *single* mind.[153]

The *Elements of Chemical Philosophy*

In 1812 Davy published what was meant to be the first volume of his *Elements of Chemical Philosophy*.[154] (As Thomas Thomson shrewdly guessed, it was also the last volume.[155]) Davy, flushed by the success and acclaim won by his researches in his years at the Royal Institution, determined to set down the essence of his natural philosophy; success relaxed the rigid discipline he had earlier imposed on himself, and he inimitably mingled fact with speculation, to the dismay of Berzelius, who failed to recognize the author of the earlier Bakerian Lectures in Davy's *Elements*.[156] Perhaps unwisely, Davy sent Berzelius a copy of the work with a conventionally polite request for criticism.[157] He received criticism, in meticulous detail, point after point, page after page.[158] Davy, fashionable hero of London science, was unused to such a response; worse still, Berzelius had shown Thomas Young his copy of the work, all crabbed over with marginal comments even more telling than those he transmitted directly to the author; and Young, in a paroxysm of tactlessness, had told Davy.[159] Thereafter, relations between Davy and Berzelius became desultory, then virtually ceased. Yet Berzelius's notes on the *Elements* are valuable for the light they shed on his and Davy's ideas. The temperamental difference between the

---

[152] R. Instn MS. 1. But Davy was not misunderstood: he had been explicit (*Phil. Trans. R. Soc.* **97**, 39, 1807).

[153] R. Instn MS. 4: lecture 2 (1812).

[154] H. Davy, op. cit. (132).

[155] *Ann. Phil.* **1**, 372, (1813).

[156] Berzelius to C. L. Berthollet, letter of autumn 1812: quoted in Söderbaum, op. cit. (129), vol. 1, p. 42.

[157] ibid., vol. 2, p. 30: 25 July 1812.

[158] ibid., vol. 2, pp. 35–59.

[159] Hartley, op. cit. (4), p. 95.

two chemists emerges clearly from Berzelius's introduction to his critical comments:

Great minds often attach too little importance to details and perhaps they are wrong to concentrate on anything other than wide and general views. Your philosophy is too far above my criticism, perhaps my head is too ill organized to be able to follow it. In hypothetical matters I limit myself to requiring of the author a precise and clear exposition of all the probabilities, and to wishing not to be prejudiced, by the expression of his opinions, in favour of any hypothesis. I feel a need to believe nothing without proof, and I want *positive* knowledge. I enjoy the study of probabilities, but wish to leave decisions about them to future research. I consider the ideas on light, caloric and electricity, as interesting fictions, which will dazzle less obstinate readers than me, and which will one day give way to the fictions of another century.[160]

Davy had a more exalted notion of the role of speculation in science; he regarded it as an essential instrument in discovery, and it figures prominently in this, his only chemical textbook.

In the introduction, he broaches the problem of affinity, and it is at once apparent that the dynamical ideas displayed in his earlier work have gained in consistency. Chemical bodies possess electrical polarity, which may be exalted by 'certain combinations' leading to chemical reaction. The phenomena of chemical separation and elective affinity occur as a result of electrical arrangements. In many cases, chemical attraction varies directly with the degree of electrical excitation by contact.[161] Such ideas as these have been adduced as evidence for Davy's acceptance of Boscovichean point atomism;[162] quite clearly, active powers are involved, and the underlying theory of matter must possess a considerable dynamical element. As Davy points out, however: 'Whether matter consists of indivisable corpuscles, or physical points endowed with attraction and repulsion, still the same conclusion may be formed concerning the powers by which they act, and the quantities in which they combine.'[163] There is no need either to reject or to retain the central corpuscle as long as one retains its central forces, for the latter are responsible for the phenomena of affinity. Once Davy had reached this point, his quest for simplicity in nature would very probably have urgently suggested the elimination of any central corpuscles, unnecessary because inactive. Davy avoided making unnecessary metaphysical pronouncements—dynamical atomism and purely dynamical point atomism would be indistinguishable in their

---

[160] See note (158) above.   [161] H. Davy, op. cit. (132), pp. 55–6.

[162] Williams, op. cit. (49), pp. 72 ff. (although it should be pointed out that Williams concedes that in 1812 Davy did not absolutely commit himself to point atomism). cf. William's essay on Boscovich and British chemists in L. L. Whyte, ed. *Roger Joseph Boscovich*, pp. 161 ff. London (1961).   [163] H. Davy, op. cit. (132), p. 56.

empirical consequences. It is, however, hard not to suspect the in- clination of Davy's sympathies.[164] Central forces, indeed, are in prin- ciple adequate to account for all the properties of matter, intimately relating the chemical, electrical, and crystallographic properties of bodies. Davy believed in this interrelation, and looked forward to its 'complete illustration' in 'the mature age of chemistry'.[165] Their re- lation was the goal towards which Davy ever strove, and provided the inspiration for Faraday's work as his successor at the Royal Institution. Combining proportions had already been found to obey quantitative laws, and electrical attractions could be measured. The presumption of a unity underlying and comprehending these super- ficially disparate phenomena made it reasonable to hope 'that at no very distant period the whole science [of chemistry] will be capable of elucidation by mathematical principles'.[166]

The more the phenomena of the universe are studied, the more distinct their connection appears, the more simple their causes, the more magnificent their design, and the more wonderful the wisdom and power of their author.[167]

The preface ends with this scientific credo, Davy's affirmation of a belief in the divine simplicity and unity of nature. The rest of the book can be seen as a protracted development of this thesis, buttressing it with fact, hypothesis, rationalization, and speculation.

A major section of the book is devoted to electrical forces in relation to chemical change. The most important aspect of electricity was prob- ably 'its connection with the chemical powers of matter, and the manner in which it modifies, exalts, or destroys these powers'.[168] This, however, did not mean that chemical and electrical powers were one and the same. Davy attempted at a stroke to distinguish and to relate these two powers, or two different manifestations of a single power: 'Electrical effects are exhibited by the same bodies, when acting as masses, which produce chemical phenomena when acting by their particles.'[169] Berzelius was baffled by this supposed distinction. What possible difference could there be between action on particles and action on the massive aggregate they constituted? He asked Davy, in what was clearly meant as a rhetorical question, 'When a force acts on each particle of a body, doesn't it have any simultaneous effect on the mass?'[170] Of course it does; yet Davy's distinction was not empty of significance. He and Berzelius were both right, within their own con- ceptual frameworks; they failed to understand one another because they understood different things by 'mass' and 'particle'. Although

---

[164] See note (99) for Oersted's dynamical atomism.
[165] H. Davy, op. cit. (132), p. 57.   [166] ibid., pp. 59–60.
[167] ibid., p. 60.   [168] ibid., p. 158.   [169] ibid., p. 164.
[170] See note (158) above.

Berzelius had encountered dynamical theories at Uppsala,[171] his reaction in this case was conditioned by atomism of the general kind
presupposed by Dalton; within such an atomistic system, his criticism
of Davy was valid. Davy, however, can have intended only a dynamical
theory in which interaction between 'particles' and 'particles', and between 'particles' and 'forces' (both concepts having 'force' in common)
meant that masses did not behave as arithmetical sums of their parts;
the relations between interlocking force fields were more complicated
and subtle than linear addition. When Berzelius formulated his own
atomic theory, with its polar atoms, he left billiard ball atomism behind; at this date, however, he himself recognized that he failed to
understand Davy because of the way the latter used his terms.[172]

The mass–particle distinction enabled Davy to declare once again
that chemical and electrical attractions were not identical. This was in
accord with his theory of voltaic action: metallic contact destroyed
electrical equilibrium, which was restored by chemical processes, both
effects co-operating in the production of continuous voltaic action.
Davy still argued, in opposition to the protagonists of the chemical and
contact theories,[173] that neither chemical reaction nor metallic contact
was in itself sufficient for the generation of voltaic electricity.

Just what was electricity? Davy recognized that

it is perhaps impossible to decide in the present imperfect state of our
knowledge. The application of electricity as an instrument of chemical
decomposition, and the study of its effects, may be carried on independent
of any hypothetical ideas concerning the origin of the phenomena; and these
ideas are dangerous only when they are confounded with facts.[174]

However, the perforation of paper by an electric discharge produced
burns on both sides, 'which may be urged as an argument against any
fluid passing through it, for it could penetrate in one direction only, and
the experiment is favourable to the idea that electricity is an exhibition
of attractive powers acting in peculiar combinations . . .'.[175] As Berzelius pointed out,[176] the results of this experiment could just as well
be used to support the hypothesis that electricity was composed of two
fluids, travelling in opposite directions. Berzelius was unable to imagine

---

[171] See Eriksson, op. cit. (39).

[172] See note (158) above. 'Perhaps the idea you had in mind may be quite correct, but
in any case the reader simply cannot deduce this idea from the *terms* you have used.'
(My italics.)

[173] M. Faraday (op. cit. (73), vol. 2, pp. 18–24) discusses the chemical and contact
theories of the origin of electricity in the voltaic pile, giving copious references to the
literature. Faraday came down heavily against the contact theory, arguing that the
contact of metals was entirely passive.

[174] H. Davy, op. cit. (132), p. 176.   [175] ibid., p. 179.

[176] Söderbaum, op. cit. (129), vol. 2, p. 44.

a force as fugitive and transitory as the electric spark, and he adopted the hypothesis of two fluids. At the end of the book, Davy reverts once more to his convictions about the fundamental simplicity of matter.[177]

Consolidation

The year 1812 marked in many ways the apex of Davy's career in fashionable society. He was knighted on 8 April, and three days later Sir Humphry married Mrs. Apreece, wealthy widow and celebrated bluestocking. Davy's life was to change; he anticipated this, and wrote to his brother John, shortly after the wedding:

If I lecture, it will be on some new series of discoveries, should it be my fortune to make them; and I give up the routine of lecturing, merely that I may have more time to pursue original enquiries, and forward more the great objects of science. This has been for some time my intention, and it has been hastened by my marriage [meaning by Lady Davy's money].[178]

Change there certainly was, though not altogether in the direction of refined scientific research. As Lord Brougham later complained, Davy committed the supreme folly of giving up his 'original natural liberal opinions for love of Lords and Ladies'.[179] This was perhaps in some measure a concession to his wife's social ambitions, eternally frustrated by her 'writhing and hard-working affectation';[180] refuge from his wife's asperity only increased Davy's inborn love of fishing, and took further toll of laboratory work. Yet this decline was gradual, and Davy still had a few years of brilliant discovery before him.

In March 1813 he appointed Faraday as his assistant in the Royal Institution, and in the autumn of that year he, together with Faraday and Lady Davy (an ill-assorted trio), left on his first continental tour. England and France were still at war, but, for all the local patriotism of scientific communities, science remained an international activity.[181]

Davy's progress through Europe was marked by repeated displays of chemical virtuosity. His investigations into the nature of iodine are an especial joy to read, for their sureness, economy, and connected reasoning[182]—guided, need it be added, by analogy.[183] Experiments on the

[177] H. Davy, op. cit. (132), p. 503.

[178] J. Davy, op. cit. (79), p. 157: H. Davy to J. Davy, spring or summer 1812.

[179] Brougham to Mary Somerville, 1834: Bodleian Library, Somerville papers.

[180] Maria Edgeworth (to Harriet), 29 March 1831: MS. in possession of Mrs. C. Colvin, Oxford.

[181] Thus in 1807 the *Institut de France* had awarded Davy the prize for the best galvanic experiments of the year—the experiments described in his first Bakerian Lecture.

[182] H. Davy, op. cit. (100), vol. 5, pp. 437–56, 456–77.

[183] cf. Siegfried, *Isis*, **54**, 247 (1963)—Davy proceeded either by *analogy*, or by *ad hoc* arguments. It should be emphasized, in agreement with Siegfried's interpretation, that Davy was an advocate of analogy in its suggestive and not in its assertive role. In 1816 he

halogens and on the nature of the diamond,[184] together with the maturation of his ideas on the nature of matter, comprised the bulk of his scientific adventures on this journey, and provided the themes for many discussions with continental scientists. Davy sought for common ground between his own work and theirs, and was disappointed when Laplace tried to moderate his eager optimism about the ultimate quantification of chemistry:

. . . November, 1813. On my speaking to him of the atomic theory in chemistry, and expressing my belief that the science would ultimately be referred to mathematical laws, similar to those which he had so profoundly and successfully established with respect to the mechanical properties of matter, he treated my idea in a tone bordering on contempt . . .[185]

Davy, no mathematician, was none the less a good Newtonian. Around this time he entered in his 'Common Place Book' the same conviction that chemistry was governed by mathematical laws, adding only that the complete exposition of chemistry would be a more *sublime* exhibition of the human mind than could be provided by mere astronomy, since chemistry was concerned with 'those molecules of matter which are invisible and of which the existence is proved by reason or aided by *experiment*'.[186]

Davy, from his earliest notebooks onwards, was capable of playing both sides in matters of speculation, of proposing two conflicting opinions—and when, as often happened, he was content to leave them in the air, there is no way of telling which notion, if either, he really favoured. However, Davy was also capable of reconciling seeming incompatibilities into a higher unity. The argument from design, coupled at times with an almost mystic sense of union with, and worship of, nature, made the synthesis and simplification of disparate ideas about the natural world his principal preoccupation. He was intent on understanding the work of 'the infinite intelligence by which so many complicated effects are produced by the most simple causes'.[187] Reduce, resolve, synthesize, simplify, unify—these were imperatives for Davy, and provided the motivation and direction for his life's work. Hence the connection between the dynamical Newtonian and the electrical theories of chemical affinity with which he juggled; they were two aspects of the same reality, which could be symbolically represented as

---

complained (op. cit. (100), vol. 5, p. 516) that '[t]he substitution of analogy for fact is the bane of chemical philosophy; the legitimate use of analogy is to connect facts together, and to guide to new experiments'.

[184] See Siegfried, *Isis*, **57**, 325–35 (1966).
[185] J. Davy, op. cit. (115), vol. 1, p. 470.
[186] R. Instn MS. 12.
[187] H. Davy, *Salmonia*, p. 86. 3rd edn, London (1832).

different facets of one geometrical figure. This explains the following notebook entry:

'What is attraction. Chemical   Electrical

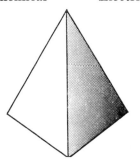

.'188

This approach enabled him, in his discussion of the nature of the diamond,[189] to connect mechanical corpuscular arrangement with optical, electrical, and chemical properties.[190]

His occasional and fluctuating adherence to Boscovich's point atomism drew its motivation from the same source—hence this entry in his 'Common Place Book':

By assuming certain molecules endowed with poles or points of attraction and repulsion as Boscovich has done and giving them gravitation and form i.e. weight measure all the phenomena of chemistry may be accounted for. They will form spherical masses when their attractions balance the repulsions and fluids or aeriform substances and chemical combination will depend upon particles meeting so as to be in polar relations so that their spheres of attraction may coincide; but we may suppose inherent powers (thus we suppose iron naturally polar with respect to the magnet)—Heat may affect chemical action by enlarging the sphere of action i.e. by expanding bodies—electrical attractions and repulsions [are] increase of primary corpuscular attractions —conductors polar/bipolar [*illegible*] nonconductors multipolar, imperfect conductors panapolar—Transparent bodies nonconductors particles further removed, and they are polar with respect to light—certain liquids imperfect conductors when solid nonconductors—This owing [either] to chrystaline arrangement which interferes with the communication of polarity. Quere cannot a voltaic apps be made of oxymuriate, or Potassa fused, or Nitre fused in Wollaston's way or is the chemical change absolutely necessary for the notion of the powers.[191]

All properties of matter are ultimately related; there is 'some relation between the primary attractive powers of the chemical elements and their electrical energies',[192] which are, in turn, connected with heat.

[188] R. Instn MS. 13j.   [189] See p. 60 below.
[190] H. Davy, op. cit. (100), vol. 5, p. 489.   [191] R. Instn MS. 12 (1814).
[192] H. Davy, op. cit. (100), vol. 6, pp. 108–9.

Such convictions led naturally to his explanation of catalytic phenomena.[193] Davy found that platinum would bring about some gas combinations at abnormally low temperatures. 'A *probable* explanation of
the phenomenon, may, I think, be founded upon the electro-chemical
hypothesis which I laid before the Royal Society in 1806; and which
has since been adopted and explained, according to their own ideas, by
different philosophers.'[194] If one supposed that oxygen and hydrogen
stood in relation to one another as negative to positive, their combination would require either the equilibration or the discharge of their
electricities.

Now platinum, palladium, and iridium are bodies very slightly positive with
respect to oxygen; and though good conductors of electricity, they are bad
conductors and radiators of heat, and supposing them in exceedingly small
masses, they offer to the gases the conducting medium necessary for carrying
off, and bringing into equilibrium their electricity without any interfering
energy, and accumulate the heat produced by this equilibrium.

As Davy explained to Faraday,[195] the heat produced, though not
sufficient to inflame the mixture, could gradually accumulate in the
wire until the gases were fired. The relation of heat to chemical affinity
was emphasized in Davy's first Presidential Address to the Royal
Society;[196] at the same time, the relation of celestial to corpuscular
mechanics, in which the weight, number, and figure of the corpuscles
were of primary importance, was stressed.

Davy, Newton's remote successor as President, was expounding
Boyle's corpuscularianism, with dynamical associations put in only as a
parenthetic afterthought, almost as a sop to his own earlier researches.
In 1821 he treated the problem of the nature of electricity as part of the
corpuscular philosophy,[197] and, although he left the solution of this
problem open, he was not unfavourable to fluid theories, or perhaps an
ether 'susceptible of electrical influences'. The dynamical theories of
Davy's most creative period now receded, not to re-emerge until his
final work, the *Consolations*, 'my six Dialogues, my legacy to the
philosophical world'.[198]

It is difficult to suggest any reason for Davy's temporary apostasy
from dynamism in chemistry. It can, in any case, have been with only
a part of his mind that he became less favourable to dynamical theories,
for the early 1820s saw Faraday's liquefaction of chlorine,[199] imme-

[193] H. Davy, op. cit. (100), vol. 6, pp. 110-12.   [194] ibid.
[195] Faraday to Benjamin Abbott, 20 Jan. 1817: R. Instn MS. 21, Warner letters.
[196] H. Davy, op. cit. (100), vol. 7, p. 11. Davy became P.R.S. on 30 Nov. 1820.
[197] ibid., vol. 6, p. 245.
[198] H. Davy to Lady Davy, 1 March, 1829: quoted in J. Davy, op. cit. (79), p. 311.
[199] *Phil. Trans. R. Soc.* 113, 160–4 (1823). *Phil. Mag.* 62, 413–15 (1823). *Ann. Phil.* 7,
89–91 (1824).

diately followed by Davy's claims: 'The experiments on the condensation of the gases were made under my direction; and I had anticipated, *theoretically*, all the results.'[200] This prediction has been ingeniously adduced as an instance of Davy's implicit acceptance of Boscovich's point atomism.[201] It is certainly true that force atomism in its most general aspect, be it Boscovichean, Kantian, or Newtonian, would indeed provide the necessary foundation for the prediction of the liquefaction of gases by pressure—if pressure is sufficiently great, it will approximate the atoms, until their outer spheres of repulsion are overcome, and attraction preponderates; the gas will then become a liquid. In these broad terms it is reasonable to accept the argument for force atomism.

It seems doubtful, however, that Davy did in fact make any such prediction, but once the fact of the liquefaction of chlorine had occurred, he was quick, as ever, to select the theory best suited to its explanation. He went on to extend this explanation by analogy, to predict the liquefaction achieved by Faraday under his direction.[202] Some evidence for this interpretation is provided by the note that Davy appended to Faraday's published paper:

In requesting Mr. Faraday to expose the Hydrate of Chlorine, to heat in a closed glass tube it occurred to me that one of three things would happen; either that it would become fluid as a hydrate, or that a decomposition of water would occur and Euchlorine and muratic acid be formed; or that the chlorine if separated would condense in the liquid form. This last result having been obtained, it evidently led to other researches of the same kind.[203]

This is already a weaker statement than the one made to Edmund; to anticipate the phenomenon as one of three possibilities is not the same thing as to predict it, particularly as only one of the three events would require force atomism for its explanation. It is, in any case, quite typical of Davy to have ready a number of alternative theories, and to select the one that fitted the facts. He did not, at this date, commit himself firmly to any one theory of matter.

Is even this weaker claim of Davy's acceptable? There would be no reason to doubt its validity, were it not that Faraday himself, and Paris, the only other witness of the experiment and in touch with both Davy and Faraday, gave somewhat different accounts of the actual course of events.

Paris told Davy of Faraday's result on the evening of 5 March 1823,

[200] Letter to Edmund Davy, 1 Sept. 1823 (my italics); cited by J. Davy, op. cit. (115), vol. 2, p. 160.
[201] Williams, op. cit. (49), pp. 129–30. The point is, however, that the choice was *not* simply one between Daltonian and Boscovichean atomism.
[202] *Phil. Trans. R. Soc.* **113**, 189–98 (1823).   [203] R. Soc. MS. P.T. 16, 16.

whereupon 'he [Davy] *appeared much surprised*; and after a few minutes of apparent abstraction, he said, "I shall enquire about this experiment tomorrow" '.[204] Paris is fallible and fond of anecdote, but he remains the only witness, both of the experiment and of Davy's reaction to the news of its outcome. Faraday accepted this account, adding that the experiment was carried out at Davy's suggestion, but 'I did not at that time know what to anticipate, for Sir Humphry Davy *had not told me his expectations*, and I had not reasoned so deeply as he appears to have done. Perhaps he left me unacquainted with them to try my ability'[205] Davy tells us that he made a practice of not telling his assistants the purpose of any experiment, so that their expectations should not prejudice their observations;[206] but his surprise at Paris's account, coupled with his own admission that the result was one of the three possibilities he envisaged, means that this cannot be used as evidence for Davy's adherence to any theory of matter or of chemical affinity.[207]

The consideration of Davy's remaining contributions to electrical theories of chemical affinity inevitably come as a postscript to a tale that has now been told. In 1823 a Committee of the Royal Society was formed for investigating the causes of the decay of copper sheathing on the hulls of ships of the Royal Navy.[208] Davy, as ever relying on the principle he had established in his first Bakerian Lecture, 'the tendency of electrical and chemical action . . . always to produce an equilibrium in the electrical powers',[209] proposed an electrochemical remedy. If the copper sheeting was rendered negatively electrical, if would not corrode.

My results are of the most beautiful and unequivocal kind; a mass of tin renders a surface of copper 200 or 300 times its own size sufficiently electrical to have no action on sea water.

I was led to this discovery by principle, as you will easily imagine; and the saving to government and the country by it will be immense . . .[210]

The principle was sound, and initial reactions were favourable; here, as in the case of the miners' safety lamp, was a demonstration of the power of scientific method in benefiting humanity. Davy was sincerely

[204] Paris, op. cit. (25), vol. 2, p. 210 (my italics).

[205] Faraday to R. Phillips, 10 May 1836: quoted by H. Bence-Jones, *The Life and Letters of Faraday*, vol. 1, pp. 375–9 (my italics). 2 vols., London (1870).

[206] H. Davy, op. cit. (100), vol. 9, p. 366.

[207] This interpretation is supported by Davy's letter to Allen, 16 March 1823 (R. Instn MS. 25, Faraday collection): 'I have lately been engaged in examining the effects of pressure in modifying the forms of bodies. Mr. Faraday in conducting some of my processes has discovered *fluid chlorine*.' Had Davy anticipated this discovery, he would surely have said as much to Allen.   [208] Paris, op. cit. (25), vol. 2, pp. 221–2.

[209] H. Davy, op. cit. (98), vol. 6, pp. 273 ff.

[210] H. Davy to John Davy, 30 Jan. 1824: quoted by J. Davy, op. cit. (113), vol. 2, pp. 175–6.

and strongly committed to this view of science.[211] Unfortunately, this time, barnacles liked Davy's remedy even better than did the Board of the Admiralty, ships' bottoms were rapidly encrusted, and Davy's answer was rejected.[212]

Then, in 1826, Davy made his last major statement on the relations of chemical and electrical changes.[213] He claimed that his theory had not altered fundamentally since his lecture 'On some Chemical Agencies of Electricity', twenty years previously. He referred to his hypothesis that chemical and electrical attraction were produced by the same cause, acting in one case on particles, in the other on masses; he also took the opportunity to state clearly his views on the value and place of hypothesis in science:

Believing that our philosophical systems are exceedingly imperfect, I never attached much importance to this hypothesis; but having formed it after a copious induction of facts, and having gained immediately by the application of it a number of practical results, and considering myself the author of it as I was of the decomposition of the alkalies, and having developed it in an elementary work, as far as the present state of chemistry seemed to allow, I have never criticised or examined the manner in which different authors have adopted or explained it,—contented, if in the hands of others it assisted the arrangement of chemistry or mineralogy, or became an instrument of discovery.[214]

Davy insisted that it was his own theory, and that it had indeed been a powerful methodological tool and instrument of discovery—these were the undeniable facts. German metaphysicians might adopt a different attitude, but they were very wrong to tell experimentalists 'that all attractions, chemical, electrical, magnetic, and gravitative, may depend upon the same cause'. The true origin of electrochemistry was the experiment of Nicholson and Carlisle, not the speculative philosophy of German Idealists.[215]

This paper exhibits Davy's much moderated dynamism of the early 1820s. Electricity no longer appears as an active force: in his tentative discussion of its nature, Davy considered only the relative advantages of the one- and two-fluid theories.[216] Point atomism is far away.

[211] This repeatedly emerges from his lectures at the Royal Institution; its practical outcome included his *Elements of Agricultural Chemistry*, London (1813); *On the Analysis of Soils as Connected with their Improvement*, London (1805); *On the Safety Lamp for Coal Mines with some Researches on Flame* (1818).

[212] Hartley, op. cit. (4), pp. 139–40.

[213] *Phil. Trans. R. Soc.* **116**, 383–422 (1826).   [214] ibid., 312.

[215] ibid., 384–6. An unfounded claim of the type that Davy is attacking here was made by Schelling in 1832, when he welcomed Faraday's production of electricity from magnetism as confirmation of *his* theory of their identity. W. Whewell, *Philosophy of the Inductive Sciences*, vol. 1, pp. 356–8, 2 vols., London (1840); Schelling, *Ueber Faraday's Neueste Entdeckung*. Munich (1832).   [216] H. Davy, op. cit. (213), 415 ff.

Consolations in Travel

There is a radical change of emphasis and style in the last of all Davy's publications, his *Consolations in Travel*.[217] The arguments are largely Neo-Platonic, though with a markedly individual flavour. Materialism is attacked, chemistry is defined deliberately without recourse to atomic hypotheses,[218] Boscovichean point atomism is at last firmly and publicly advocated,[219] and chemical attractions are described as being peculiar forms of electrical attractions. The arguments for these positions are presented cogently, but wrapped around with speculations on religion and philosophy. The result is a unique and puzzling amalgam—but it does offer many keys to Davy's mind, and not all of these are illusory.

It is likely that the *Consolations*, together with the unpublished 'Dialogues' in the Royal Institution, represent the final undisciplined release of metaphysical predilections that Davy the scientist had found attractive but had ruthlessly repressed and kept out of his laboratory.[220] Davy's metaphysics was firmly isolated from his experimental philosophy, and this dualism was conscious and intentional. His whole career exemplifies this separation; but it is important to distinguish natural theology, religion in relation to the natural world, from metaphysical speculations in general. Religious beliefs, to those who hold them, are in a class apart from mere metaphysical speculations: they are certainties, absolute and irrevocable. Charles Darwin quotes, in his autobiography, a lady who told his father, 'Doctor, I know that sugar is sweet in my mouth, and I know that my Redeemer liveth.'[221] Just so did Davy know that the world, designed and created by God, was ultimately one, simple and purposeful. He did not allow his religion to trespass into the laboratory, but he did let it guide and determine his fundamental beliefs about nature; and although he would not allow religion to decide upon the content of a theory, nevertheless he could not but let it be effective in the selection of theories.[222]

Davy's natural theology was basic to his philosophy of science; it appears as basic also to the *Consolations*, which are his personal religious statement, concerned with his own salvation through the

[217] A good brief account of this work is given by D. M. Knight, 'The scientist as sage.' *Studies in Romanticism*, **6**, 65–88 (1967).

[218] A. Davy, op. cit. (100), 9, pp. 362–3.

[219] ibid., vol. 9, p. 388. Davy's reference to Mako for information about point atomism is dealt with in the section on 'matter' below.

[220] cf. Siegfried, 'Davy's *Consolations in Travel*': short paper read at Br. Soc. Hist. Sci. meeting, Jan. 1968.

[221] *The Autobiography of Charles Darwin*, ed. N. Barlow, p. 96, London (1958).

[222] cf. T. H. Levere, 'Faraday, Matter, and Natural Theology', *Br. J. Hist. Sci.*, 95–107 (1968).

acquirement of and search for natural knowledge. Davy's natural philosophy is tied to his theology of nature by his belief that

whenever we attempt metaphysical speculations we must begin with a foundation of faith. And, being sure from relevation, that God is omnipotent and omnipresent, it appears to me no improper use of our faculties, to trace *even in the natural universe*, the acts of his power and the results of his wisdom, and to draw parallels from the infinite to the finite mind.[223]

In the *Consolations*, Davy sets out from two basic points, secure in his faith that the paths leading from them are convergent.[224] The basic teachings of Christianity provide him with one starting point[225] —the belief that an omnipotent, omniscient, and benevolent God sustains the world, having created it from nothing, and made man in his own image, giving him an immortal soul, a soul redeemed by Christ. Davy, however, does more in the *Consolations* than state religious ideas. There is another starting point and source for his 'Dialogues': the latest thoroughly confirmed empirical conclusions of science—and science, Davy tells us through the mouth of one of his characters, 'is in fact nothing more than the refinement of common sense making use of facts already known to acquire new facts'.[226] The most recent developments in contemporary chemistry, physics, and geology are brought forward as the 'Dialogues' progress. This is only to be expected from Davy.

The interplay between religion and science generates Davy's philosophy, which enters the book only because Christianity and science are there already. If this interpretation is correct, then Davy's philosophy, tinged with Neo-Platonism,[227] is in part a direct result of his Christianity. The Neo-Platonism rampant in the early years of the church found its way into the very language of the bible;[228] a well-read Christian of Davy's generation would be saturated in this mode of thought, passed down to him by English writers in the broadly Christian Platonic tradition.[229]

---

[223] H. Davy, op cit. (100), vol. 9, p. 381   [224] See ibid., p. 270.
[225] This is clear throughout the work—pp. 263–77.
[226] ibid., p. 355.
[227] The ascription of neo-Platonic notions to Davy is an over-simplification necessitated by brevity. Davy's awareness of seventeenth-century Cambridge neo-Platonism, perhaps owed to Coleridge, was reinforced emotionally by the Pantheism evident in his poetry (see, for example, Davy's *Works*, vol. 1, *passim*); but Pantheism was intellectually incompatible with Christianity, so that emotion and intellect were here at odds. For a detailed discussion of this clash, suffered and clearly perceived by Coleridge, the reader is referred to T. McFarland, *Coleridge and the Pantheist Tradition* (Oxford, 1969); reference is there made to Davy, pp. 87, 165, whose case I believe to have been similar.
[228] e.g. John 1: 1. 'In the beginning was the Word, and the Word was with God, and the Word was God.'
[229] E.g. H. More, R. Cudworth.

Davy was most certainly well-read. The *Consolations* bear witness to the astonishing catholicity of his knowledge, and make it impossible to rule out anything as a conceivable influence on his intellectual development.

The problem of Davy's philosophical loyalties seems, therefore, to be artificial. If, however, as I believe, his preference was for British over German philosophy, the *Consolations* suggest that it was not because the former eschewed, as the latter did not, system spinning and transcendental sophistication. It was rather because Davy was English, and German philosophy had little to tell him, on what he considered were crucial topics, that he would not have read already in earlier native authors.[230]

Davy's theory of matter

A principal contention of this chapter has been that, on the assumption of a dynamical theory of matter, ideas about chemical affinity are subsumed under the heading 'nature of matter'. Davy's atomism was generally within the British tradition of Newtonianism in its dynamical aspects; his conviction of the fundamental simplicity and unity of matter, together with his conception of chemical affinity as one of the active powers of matter, agrees with any theory that assumes active powers inherent in matter, together with fundamental simplicity of that matter, be it corpuscular or purely dynamical. Davy's deliberate vagueness protects him from the criticisms that could be levelled at a more specific theory, and renders it difficult to ascribe sources to any of Davy's ideas. His own eclectic ability further complicates the exercise. His unpublished 'Dialogues' furnish many examples of this: for example, next to passages of the purest dynamism, with almost animist undertones, one suddenly finds the suggestion that 'Possibly the true solution of all the phenomena of chemistry may ultimately be mechanical or dependent upon the primary attractive powers of the corpuscles of matter influenced by their forms and masses . . .'.[231]

Davy, believing that the natural world was a divinely ordained unity, intended no contradiction between these seemingly antagonistic viewpoints; to state the paradox baldly, atomism was one possible aspect of dynamism. Thus although Davy repeatedly referred to Boscovich's

---

[230] In discussing Davy's philosophical leanings, I should also mention his involvement in the early nineteenth-century debate, Christianity versus Pantheism. Davy had at least a youthful flirtation with Spinozist Pantheism; see C. Carlyon, *Early Years and Late Reflections*, pp. 193 ff. (London, 1856), and T. McFarland, op. cit. (227), p. 87. A Romantic sympathy with nature is evident in much of Davy's poetry. This is, of course, extremely peripheral to any discussion of Davy as a natural philosopher.

[231] R. Instn MS. 9.

point atomism, it is not safe to assume that he unquestioningly adhered to it.[232]

In the *Consolations*,[233] the Unknown, generally Davy's own mouth-piece, explains that 'as we can never see the elementary particles of bodies, our reasoning upon them must be founded upon analogies derived from mechanics, and the idea that small indivisible particles follow the same laws of motion as the masses which they compose'. This line of argument is pure Newton and is, not unreasonably, taken to oppose point atomism, and the Unknown is referred to 'the hypo-thesis of Boscovitch, which is well explained in the *Institutio Physica* of Mako'.[234, 235]

The Unknown, of course, knows all about this, and explains that his analogical arguments from mechanism are entirely compatible with point atomism. He even adduces experimental evidence to suggest that the minimal assumptions of point atomism suffice to account for the phenomena. He does this, one must admit, by a strange use of terms.

In the *Consolations*, Davy seems to have adopted point atomism, and rejected the dynamical atomism he had once espoused. He had known of Boscovich's theory for years[236]—while he worked at the Royal Institution, the 1763 edition of Boscovich's *Theoria* was acquired for the library;[237] his predecessor, Garnett, had included a brief account of the theory in his lectures at the Royal Institution,[238] and there were many other secondary sources. In addition to Mako's account, there were

[232] The main references to Boscovichean point atomism in Davy's work are:

Elements of chemical philosophy, pp. 56, 489, London (1812).
'Common Place Book', entry of [1814]: R. Instn MS. 12.
Collected Works, vol. 7, p. 102. 9 vols. London (1839): Royal Society 1826 Anniversary Meeting.
ibid., vol. 9, p. 388.

[233] H. Davy, op. cit. (100), vol. 9, pp. 387–8; this is missing from separate editions of the *Consolations in Travel*.

[234] ibid., vol 9, p. 387.

[235] P. Mako, *Compendiaria physicae institutio quam in usum auditorum philosophiae e lucubratus est*, part 1. 2nd edn, Vindobone (1766). It is tantalizing not to know at what stage of his career Davy read Mako. Boscovich as a Newtonian, a dynamical theory of matter that explains the diversity of chemical phenomena in terms of point atoms and fundamental simplicity, the use of dynamical atomism to account for precipitation, fermentation, etc., the small number of true chemical elements—these and many more notions that occur in Davy's writings occur also in Mako's text (pp. A5, 1, 3–4, 54 ff.). If Davy read the book in his youth, it may have been very important in his development; however, his notes give no indication of this. If Davy read the book in the latter part of his life, he would have been struck by a number of familiar and fruitful ideas. But why, when giving people a reference to Boscovich's theory of matter, offer so obscure a work? Perhaps Davy was merely showing off his erudition.

[236] At least since he made the entry in his 'Common Place Book' on his first continental tour.

[237] It is not in the 1809 catalogue, but appears in the 1821 catalogue. Probably Davy's Latin was good enough to use it; cf. J. Davy, op. cit. (115), vol. 1, p. 15.

[238] D. M. Knight, *Atoms and Elements*, p. 38. London (1967).

A M—F

comments in Young's *Lectures*[239] and Thomson's *System of Chemistry*,[240] longer accounts in the *Encyclopaedia Britannica*,[241] and a full hundred pages devoted to the topic of Robison's popular *System of Mechanical Philosophy*.[242] Acquaintance, however, is simply not enough. Evidence can be produced that Davy knew of and used various other theories— notably Newton's;[243] he *used* them all, and transformed them in use. Only in this sense can one claim that Davy was a Boscovichean in the years of his creative research. Davy's oscillation between different kinds of dynamism—from dynamical atomism to pure force theories— cannot be explained precisely by saying that his attitude to theories was heuristic; however, his inconstancy does become thereby less of a puzzle.

Arrangement and chemical affinity

Chemical combination arises from the exertion of forces; their nature, degree, and disposition are all germane to its action. Degrees of chemical affinity have been hinted at in Davy's electrochemical arrangements of the elements; and his correlation of their electrical with their chemical properties, within a general theory of matter, has explained the nature of chemical affinity as Davy understood it. There remains the part played by the disposition of the forces associated with matter; it is time to consider Davy's solution of this aspect of the problem.

In 1801 Coleridge wrote to Davy, sadly complaining that 'That which must discourage us in it [i.e. chemistry] is, that I find all *power* of vital attributes to depend on modes of *arrangement*, and that chemistry throws not even a distant rushlight glimmer upon this subject.'[244] Davy shared Coleridge's conviction of the importance of arrangement for a full understanding of chemistry, but not his pessimism. Newton had given his authority to the 'sublime chemical speculation'[245] that

---

[239] T. Young, op. cit. (113), vol. 1, p. 614. It is interesting that Young provides an illustration of R. Harre's thesis in *The Anticipation of Nature* that it is impossible to arrive at a generally acceptable criterion of simplicity. Young says that Boscovich's theory 'has prevailed very widely among algebraical philosophers . . . Such methods may often be of temporary advantage . . . but the grand scheme of the universe must surely, amidst all the stupendous diversity of parts, preserve a more dignified simplicity of plan and of principles, than is compatible with these complicated suppositions'. Young was a good enough mathematician to realize the practical complexities of the application of Boscovich's method; Davy, no mathematician, only regarded the simplicity of the premises.

[240] T. Thomson, *A System of Chemistry*, vol. 3, pp. 598–9. 4th edn, Edinburgh (1810), also various other editions (see bibliography).

[241] e.g. *Encycl. Brit.* vol. 4, pp. 41–59, art. *Boscovich*. 4th edn (1810).

[242] J. Robison, *A System of Mechanical Philosophy*, ed. Brewster, pp. 267–368. 4 vols., Edinburgh (1822).   [243] Hence this chapter.

[244] 4 May 1801: quoted by J. Davy, op. cit. (79), p. 90.

[245] H. Davy, op. cit. (132), p. 489.

the forms of different bodies might depend upon different arrangements of the same particles of matter, and Priestley had been concerned with the internal structure of bodies, as revealed by the application of electrical powers to chemistry. In one of his first lectures at the Royal Institution, Davy announced prematurely that 'The researches of modern chemistry . . . have demonstrated that all natural bodies consist of different arrangements or combinations of a few simple parts or elements.'[246] In spite of this confidence, repeatedly expressed, Davy was not yet ready to commit himself to the notion that the different qualities of different chemical substances depended upon arrangement, rather than upon the chemical natures of the individual constituents.[247] In his first Bakerian Lecture, he stressed the importance of electrical forces in arranging the salts in solution, prior to their decomposition. In 1810 he wondered whether 'the faculty of combination of Bodies be owing to a *peculiar arrangement* of their parts, or to powers which are exhibited by electrical energies',[248] and thereafter he appeared to identify different electrical states with different arrangements.

Davy's subsequent work was influenced at every turn by his ideas about arrangement. Such ideas underlay his work on the halogens and on diamond.

It was from dissatisfaction with Lavoisier's table of elements that Davy progressed to the discovery of the halogens.[249] Since these could not be decomposed, they must be regarded empirically as elements; therefore the halogen acids, previously assumed to contain oxygen, the principle of acidity, could not contain oxygen. In any case, the idea that oxygen was the principle of acidity had been seriously weakened by Davy's own discovery of the alkali metals; caustic soda and caustic potash both contained oxygen—perhaps this might be also the principle of alkalinity? Or perhaps there was no such thing, and, as Davy came to believe, acidity and alkalinity depended 'upon peculiar combinations of matter, and not on any peculiar elementary principle'.[250]

Davy's work on the diamond has recently been presented as the thread connecting all his major work in the years 1808–14.[251] In 1809 he attempted to analyse diamond using potassium. He thought that the potassium was partly oxidized in the process, and suspected that diamond contained some oxygen. This led to his attempt to synthesize diamond by oxidizing charcoal: he heated the latter to white heat in oxymuriatic acid gas (chlorine), expecting it to yield its oxygen to the charcoal. His expectation, however, was not fulfilled, and this failure

[246] J. Davy, op. cit. (115), vol. 1, p. 171.
[247] H. Davy, op. cit. (100), vol. 5, pp. 170 ff.
[248] R. Instn MS. 3, lecture 6 (my italics).   [249] Siegfried, *Isis*, **57**, 325–35 (1966).
[250] H. Davy, op. cit. (100), 5, p. 456.   [251] Siegfried, op. cit. (249).

led to his investigation of oxymuriatic acid, and to the conclusion that it contained no oxygen, but was instead an element sharing many properties with oxygen. Later, and independently, came the work on fluorine, and the recognition of the existence of a whole new class of elements, the halogens. Davy suspected that there might be other, as yet undiscovered, bodies belonging to this class.

The conjecture appears worth hazarding, whether the carbonaceous matter in the diamond may not be united to an extremely light and subtle principle of this kind, which has hitherto excaped detection, but which may be expelled, or newly combined, during its combustion with oxygen. That some difference must exist between the hardest and most beautiful of the gems and charcoal, between a non-conductor and a conductor of electricity, it is scarcely possible, notwithstanding the elaborate experiments that have been made on the subject, to doubt: and it seems reasonable to expect, that a very refined or perfect chemistry will confirm the analogies of nature, and shew that bodies cannot be the same in composition or chemical nature, and yet totally different in all their chemical properties.[252]

Yet when he burned diamond in Florence, he was forced to conclude that it was pure carbon, containing no oxygen; the difference between it and charcoal resided therefore necessarily in its crystalline state—in its mode of arrangement. Davy must have welcomed his conclusion, for it increased the unity, and therefore the simplicity, of knowledge about the natural world. 'Transparent solid bodies are in general non-conductors of electricity, and it is probable that the same corpuscular arrangements which give to matter the power of transmitting and polarizing light, are likewise connected with its relations to electricity . . .'[253]

Arrangement was to serve as the master-plan for the coordination and union of the chemical world. The spatial relations of forces, of parts of matter, would determine all their properties, physical and chemical: like Boscovich's interlocking matrices of power, Davy's arrangements would produce or prevent chemical combination, and in this sense arrangement constituted chemical affinity. An undated note for a lecture at the Royal Institution reads:

It seems probable from the past progress of discovery—That Chemistry at no very distant period will like *astronomy* become a science founded upon mathematical principles—That the number of elements will be diminished: and that arrangements of a very simple nature will explain those phenomena which are now referred to complicated and diversified agents. I shall shew in the conclusion of the course that if it be assumed the two electrical powers or substances are capable of producing different bodies by acting upon or combining with the same species of matter—The results must be the

[252] H. Davy, op. cit. (100), vol. 5, pp. 435–6.   [253] ibid., vol. 5, p. 489.

types of all combinations, and in such an hypothesis . . . the regularity of the law of proportions must be the necessary consequence of the powers by which they are produced.

Nature infinitely complicated in the minute details of her operations when well investigated is always found wonderfully simple in the grand mechanism of her works.—

The uniformity of the succession of events in our globe, the constant decay, and constant renovation of the forms of things—the infinite mutations of the parts of matter—the conservation of the order of the whole demonstrate at once unity of design and unity of power.—

The different arrangements may be compared to the characters said to be inscribed upon the leaves scattered abroad by the Sybils which separately had no signification but which when connected in their proper order became not merely [an] intelligible but likewise a divine language.[254]—or like the parts of a melody in music their efforts depend upon their relations and connection and when examined as a system, they appear as sounds of one voice, impulses of one eternal intelligence.[255]

In this superb passage, Davy has brought his thoughts on arrangement, matter, electricity, order, unity, simplicity—and affinity—into a relation as complete and as unified as his natural theology. This is truly his philosophy of science, his legacy and statement of faith to the scientific world: and here let us leave him.

[254] cf. Boscovich, *Theoria philosophiae naturalis*, para 541 (Venice, 1763 edn), trans. J. M. Child. Chicago (1922).

[255] R. Instn MS. 1.

3

FARADAY

W E have seen that Davy's was a dynamical chemistry, and that, in a sense, Coleridge was right in saying that his friend's major discoveries were 'made during the *suspension* of the mechanic Philosophy related to chemical Theory'.[1] In one curious dialogue in the Royal Institution, Davy even compared man to a point atom, and active powers to God's creative energy.[2] The simile is revealing: his natural theology underlay his dynamical picture of the physical world, and these two aspects of his thought were mutually supporting. Faraday inherited, modified, and extended this combination, just as he inherited, modified, and extended Davy's ideas on chemical affinity.

Faraday's apprenticeship—dynamical Newtonianism

On 1 March 1813 Faraday was appointed as Davy's assistant at the Royal Institution. He came with the most ardent enthusiasm to emulate his new master. Less than a year before, he had attended one of Davy's lectures on chlorine, and poured out his excitement to Benjamin Abbott, a friend with whom he corresponded for mutual improvement.

I have again gone over your letter, but am so blinded that I cannot see any subject except chlorine to write on; . . . be not surprised, my dear A., at the ardour with which I have embraced the new theory. I have seen Davy himself support it. I have seen him exhibit experiments, conclusive experiments, explanatory of it, and I have heard him apply those experiments to the theory, and explain and enforce them in (to me) an irresistible manner. Conviction, Sir, struck me, and I was forced to believe him; and with that belief came admiration.[3]

Shortly after Faraday became Davy's disciple, the two of them, together with Lady Davy, left on their first continental tour. Lady Davy's malicious tongue was a sore trial to the party, especially to Faraday, whose pride would not succumb to her mortifications; there is no pride so strong as that of a humble man. The journey was, how-

---

[1] S. T. Coleridge to Lord Liverpool, 28 July 1817: quoted by Griggs, ed. *Collected Letters of S. T. Coleridge*, vol. 3, p. 760. 4 vols., Oxford (1956–9).

[2] R. Instn MS. 9.

[3] R. Instn Warner MSS. letter of 19 Aug. 1812: quoted by H. Bence-Jones, ed. *The life and letters of Faraday*, vol. 1, pp. 30–1. 2 vols., London (1870).

ever, very valuable in Faraday's education. Davy's exemplary investigations on iodine, on the oxides of chlorine, and on diamond, gave Faraday models of scientific experiments, and showed him the power for discovery of analogical reasoning and hypothesis in science.[4] Throughout their travels, Davy, 'as is the practice with him, goes on discovering'.[5] Even more important was Davy's conversation, 'a mine inexhaustible of knowledge and improvement'.[6] Faraday was decidedly an admiring and attentive pupil. The experiments on the combustion of diamond led Davy to conclude that chemical and physical properties might both be functions of arrangement; the lesson was not lost upon Faraday. It was on this journey, too, that Davy made in his 'Common Place Book' a lengthy entry on the explanatory powers of the theory of point atomism,[7] and it is possible that he discussed the hypothesis with Faraday. At any rate, when they returned to England, and Faraday began lecturing at 'Mr. Tatum's', the City Philosophical Society,[8] he did suggest that Boscovichean point atomism might serve as an alternative to material solidity—'but there is reason to believe that the primary atoms of matter are indivisible, and unalterable'.[9]

His youthful adoption of Davy's ideas covered the whole field of science; a student has to learn his master's ideas before he can judge them and form his own. Whereas he had previously been exceedingly well pleased with Thomson's definition of chemistry as being the science of insensible motions,[10] he now deliberately redefined it as knowledge of the powers and properties of matter.[11] At this early date he was already arguing that matter is known by its properties and powers alone, and that it is otherwise inconceivable;[12] in 1844 he was to

[4] Years later, he was to write: 'You can hardly imagine how I am struggling to exert my poetical ideas just now for the discovery of analogies and remote figures respecting the earth, sun, and all sorts of things—for I think that is the true way (corrected by judgment) to work out a discovery.' (Faraday to Schoenbein, 13 Nov. 1845: quoted in G. W. A. Kahlbaum and F. N. Derbyshire, ed. *Letters of Faraday and Schoenbein*, p. 149. Basle and London (1899). cf. H. Davy, R. Instn MS. 9 (and *Consolations in Travel*): 'The imagination must be active and brilliant in seeking analogies yet entirely under the influence of the judgment in applying them.'

[5] Faraday to Huxtable, 13 Feb. 1815 from Rome: quoted by Bence-Jones, op. cit. (3), vol. 1, p. 195.

[6] Faraday to R. G. Abbott, 6 Aug. 1814: quoted by Bence-Jones, op. cit. (3), vol. 1, p. 147.　　　　　　　　　　　[7] R. Instn MS. 12; see p. 55 above.

[8] The City Philosophical Society was founded by Mr. Tatum in 1808. Tatum introduced Faraday as a member in 1813, when the Society 'consisted of thirty or forty individuals, perhaps all in the humble or moderate rank of life. Those persons met every Wednesday evening for mutual instruction . . . This Society was very moderate in its pretensions, and most valuable to the members in its results.' (Bence-Jones, op. cit. (3), vol. 1, p. 58.) Faraday gave 17 lectures there between January 1816 and 1818. (Instn elect. Engrs MS.)

[9] Instn elect. Engrs MS. 'Lecture Book', pp. 4–5.

[10] Faraday to B. Abbott, 11 Aug. 1812: quoted by Bence-Jones, op. cit. (3), vol. 1, p. 29.　　　[11] ibid., p. 2.　　　　[12] ibid., pp. 1–2.

use just this argument to support the hypothesis of point atomism,[13] even though it postulated entities that, correctly understood, were literally inconceivable.[14]

In his first lecture, Faraday stated his guiding chemical philosophy, which remained unchanged throughout his career. First, he made it quite clear that his concern 'is not so much the particular, as the general laws of the science'. Broad views and comprehensive theories were always his object; at a time when scientists tended to be either chemists or physicists, he, like Davy before him, was a natural philosopher, who regarded the sciences as ultimately one. The powers of matter were the web uniting the sciences, and serving for their construction and object. He saw chemistry as part of this web:

> The science of chemistry is founded upon the cohesion of matter, and the affinities of bodies; and every case, either of cohesion or of affinity, is also a case of attraction. It is, therefore, of the utmost importance that we should become acquainted with attraction in general before we descend to particular instances. When, also, I have informed you that the powers which cause the cohesion of similar matter, and the combination of dissimilar matter, are actually those which we have been considering under the names of attraction of aggregation and electrical attraction, it will immediately be seen that I have not entered into a detail of irrelevant and superfluous matter, but have been employed in giving first principles for the consideration of powers immediately productive of chemical phenomena. That the attraction of aggregation and chemical affinity is actually the same as the attraction of gravitation and electrical attraction I will not positively affirm, but I believe they are.[15]

The simplicity criterion is also invoked.

> The means employed by nature to obtain her ends are admirable and well calculated to excite our astonishment by their simplicity efficacy and perfection. A few general laws hold the universe together and a few variations and combinations of those laws impressed upon matter form all its varieties and give rise to those diversified phenomena which constitute the present state of things.[16]

From the anisotropy of crystal form he deduced anisotropy of atomic powers, manifested as a polarity of attraction and repulsion in their particles. This was hypothetical but it was nevertheless very useful, 'for it enables us to arrange a number of facts which before were insulated and to substitute order and regularity for complexity and confusion'.[17]

The foregoing extracts show that Faraday regarded the true ex-

---

[13] *Phil. Mag.* **24**, 136–44 (1844); see pp. 99–102 below.
[14] Hence, for example, Daubeny's reaction (*Introduction to the Atomic Theory*, p. 47 (Oxford, 1850)): 'I cannot bring myself to assign any properties at all to mere space.'
[15] Instn elect. Engrs MS. 'Lecture Book', pp. 30–1.
[16] ibid., 2nd lecture, p. 40.   [17] ibid., p. 71.

planation of chemical affinity as being the key to a correct understanding of chemistry. Accordingly, in this, his first lecture course, he referred to it constantly, and devoted an entire lecture to the topic.[18] Like Davy, he revered Newton and considered that it was he who had

first introduced correct notions into this part of Science and his opinions [that attraction was responsible for solution and for reaction] form the basis of the modern theory . . . His principles on this subject have been universally adopted [sic] and have also been refined and rendered more definite. They are those to which I shall adhere in this lecture.[19]

However, he showed himself more a disciple of Davy than of Newton in stressing the modification of the powers of matter that could occur on combination. The greater the power with which bodies combined, the greater the change in their properties.[20] He likewise accepted the conclusions Davy had drawn about the close correlation between electrical powers and chemical affinity, which would justify the classification of bodies in terms of 'their electrical energies or their respective inherent states'.[21]

Faraday, in concluding these lectures, stressed that practical and applied science was raised upon the foundation of elementary chemistry, by which he meant the 'laws and forces on which the science is founded'.[22] Chemical affinity and the laws governing its operation were the foundations upon which the whole edifice of chemical science was raised.

Faraday's conviction that the powers of matter were subject to modification underlay his work on the affinities between carbon and chlorine. These elements appeared unwilling to combine directly, even under abnormally active conditions. Davy's experiments on the nature of diamond had included an unsuccessful attempt to burn diamonds in chlorine gas.[23] One may conjecture that Faraday's supposed adherence to point atomism persuaded him that carbon and chlorine must have an affinity for one another. The formation of ethylene dichloride could then be explained by assuming that hydrogen served as an intermedium, facilitating the operation of the mutual affinity between the other two constituent elements. There are, however, two other possible interpretations, which, at this date, carry more conviction. First, Faraday was an avowed Newtonian; Newton regarded attraction as a general property of matter—as Faraday understood him, attraction was an inherent property. Newton's holding the contrary opinion is irrelevant: Faraday specifically tells us that this is how he understood Newtonian attraction.[24] If it was a property of all matter, then all

[18] ibid., pp. 75–109.   [19] ibid., p. 77.   [20] ibid., p. 89.   [21] ibid., pp. 97–104.
[22] ibid., lecture 17: conclusion quoted by Bence-Jones, op. cit. (3), vol. 1, pp. 258–61.
[23] Faraday, 'Common Place Book', entry for 28 March 1814.
[24] Instn elect. Engrs MS. Lecture Book, pp. 77–81.

chemical substances should have an attraction or affinity for one another. Although this might be masked or temporarily overcome—for example, competing attractions could nullify its effects—it would surely manifest itself under favourable circumstances.[25] The second and probably complementary reason for Faraday's thinking it possible to obtain a compound or compounds of carbon and chlorine was that Davy believed in the possibility. We have already seen that Faraday tended to accept Davy's speculative hypotheses as being probably true.

One of Davy's earliest notebooks at the Royal Institution includes the entry 'Mem. to try the Combustion of charcoal in Muriatic Acid by the galvanic spark'.[26] This was written in hopeful anticipation of the decomposition of oxymuriatic acid by carbon; but in 1810 when Davy was beginning to suspect that oxymuriatic acid gas did not contain oxygen, he still thought that there was reason for believing that the combination of carbon and chlorine could be effected 'by the intermedium of hydrogen'.[27] Two years later his brother John wrote a paper 'On a gaseous Compound of Carbonic Oxide and Chlorine',[28] and then in 1814 came Humphry's attempt to burn diamond in chlorine. Some sort of force atomism underlay Davy's conviction: naïve Newtonianism would not account for his idea that hydrogen might act as the intermedium for the required combination. It is, however, unnecessary to invoke the same hypothesis for Faraday, who as late as 1823 admitted that his liquefaction of chlorine came as a complete surprise to him;[29] his theories were not yet held firmly enough to have predictive value, although they served for amusement and speculation.[30]

[25] Instn elect. Engrs MS. Lecture Book, p. 81.   [26] R. Instn MS. 13c.

[27] H. Davy, *Collected Works*, ed. J. Davy, vol. 5, p. 300. 9 vols., London (1839).

[28] Carbonyl chloride; *Phil. Trans. R. Soc.* **102**, 144–58 (1812).

[29] There is however an undated entry on p. 10 of Faraday's MS. book 'Chemical Notes' (Instn elect. Engrs) in which he wonders about the 'General effects of compression either in condensing gases or producing solutions or even causing combination at comparatively low temperature . . .

Ascertain effects of compression

,,   ,,   ,,   ,,   with solvent powers

,,   ,,   ,,   ,,   ,,   ,,   ,,   & cold.'

The difficulty is that one cannot be sure how accurate an indication is provided by the queries in this volume of Faraday's theoretical opinions as opposed to his speculations. Not all the entries presuppose a dynamical theory of matter, and we have Faraday's statement of his surprise at the liquefaction of chlorine to suggest that such theories were not strongly held before the liquefaction. Perhaps the only course is to indicate the possibilities, while refraining from committing oneself to any particular brand of dynamism.

[30] Faraday's 'Common Place Book' includes numerous excruciating puns, different ways of spelling his name, a chemical love-letter, and the entry (p. 324):

'Questions for Dorset Street,

An experimental agitation of the question of electrical induction.

"Bodies do not act where they are not." Query is not the reverse of this true? Do not all bodies act where they are not; and do any of them act where they are?'

In a lecture given in 1817 to the City Philosophical Society, Faraday admitted that there was no known binary compound of carbon and chlorine, but he offered 'some proofs of the attraction between charcoal and chlorine'. His 'proofs' consisted in pointing to the existence of phosgene gas and of the Dutch Chemists' oil.[31] The formation of the latter was 'a direct combination of the elements, and not a mere case of union between chlorine and olefiant gas'.[32] His 'Common Place Book' shows that he worried away at the problem in the winter of 1817; then comes a note, 'Chlorine and Carbon—made out autumn of 1820'.[33] In October 1820 Humphry Davy wrote to his brother, 'Faraday has discovered a combination of Chlorine and charcoal (not to be mentioned)'.[34] By December, Faraday was able to announce the discovery of two new compounds of chlorine and carbon, and also of a compound of iodine, carbon, and hydrogen.[35] The discovery of yet another compound of chlorine and carbon soon followed.[36] Faraday tells us that it was Davy's reasoning on the 'triple compound of chlorine, carbon, and hydrogen' that had given him the confidence to search for these substances.[37] Belief in the existence of an affinity between two elements had led to the discovery of their compounds, and helped to focus Faraday's attention on the nature and patterns of molecular and atomic forces.

We have now encountered Faraday's dynamical Newtonianism, with his concern for the powers of matter and the corresponding importance that he attached to chemical affinity. His early ideas on the nature of affinity have been briefly indicated, but, as was the case with Davy's theories, the aspect of affinity comprised under the heading 'arrangement' or 'structure' was also of central importance, and even a superficial survey would be incomplete if it did not take this into account. The unity of forces both chemical and physical, and the spatial arrangement of matter with its powers, were the twin keys to Faraday's proposed integration of chemical with physical science.

On 10 September 1822 he entered in his laboratory diary:

Polarized a ray of lamp light by reflection and endeavoured to ascertain whether any depolarizing action exerted on it by water placed between the poles of a voltaic battery in a glass cistern—one Wollaston's trough used— the fluids decomposed were pure water—weak solution of sulphate of soda

[31] Ethylene dichloride; 'Lecture Book', pp. 353 ff.

[32] *Q. J. Sci. Arts*, **6**, 358–60 (1819).

[33] 'Common Place Book', p. 167.   [34] R. Instn MS. 26.

[35] C_2Cl_6, C_2Cl_4, $C_2H_4I_2$; *Phil. Trans. R. Soc.* **111**, 47–74 (1821). See L. P. Williams, *Michael Faraday*, pp. 121 ff., 123. Chapman and Hall, London (1965).

[36] Hexachlorobenzene; M. Faraday and R. Phillips, *Phil. Trans. R. Soc.*, **111**, 392–7 (1821).

[37] *Phil. Trans. R. Soc.* **111**, 48 (1821).

and strong sulphuric acid. None of them had any effect on the polarized light either when out of or in the voltaic circuit, so that no particular arrangement of particles could be ascertained in this way.[38]

If Faraday had succeeded in discovering the arrangement of material particles that he sought in this experiment, he would at one stroke have reached most of the fundamental conclusions that his life's work was devoted to establishing. The very fact that he could set up the experiment indicates how wide and yet how unified were his presuppositions about the forces in nature, for he clearly hoped to show that light, electrical and chemical forces, and mechanical arrangement were all comprised in a single unity. Negative results never discouraged him, and over the next forty years his experiments, successful and unsuccessful, were all motivated by the same hope of discovering unity in nature by correlating diverse phenomena. Reasoning, one may suppose, from the shape of crystals to their inner arrangement, Faraday concluded that crystalline forces were anisotropic, and assumed that the regular alignment of particles would lead to a directional force, so that crystals would have axes of preferred alignment with one another. In October 1824 he tested this hypothesis, and failed to arrive at any positive result.[39] He was not thereby dismayed, nor did he alter his convictions about the general nature and interrelation of physical forces. Examples of Faraday's continuous search for a unity underlying all the apparent diversity of nature could be endlessly multiplied,[40] and in each case one would be forced to notice that failure acted only as a spur to further and more intense investigation and search for that unity whereof he never despaired, and whose existence, indeed, he did more than any other single man to demonstrate.

Chemical affinity is a guiding thread through the convolutions of Faraday's science. All natural forces, however, were subsumed in his mind under a single principle, favouring the dynamism he came to adopt in his theory of matter. An anticipation of the argument at this stage will help to confer the same unity on the chapter that natural theology promoted in his science.[41] For this purpose we shall assume that Faraday's theory of matter has already progressed (as it was later to do) to force atomism;[42] within this framework, theories of affinity

---

[38] M. Faraday, *Diary*, ed. T. Martin, vol. 1, p. 71. 7 vols. and index, London (1932–6).
[39] ibid., vol. 1, p. 167, para. 1824.
[40] For a single example, see Faraday, op. cit. (38), vol. 1, pp. 309–10.
[41] R. E. D. Clark has come independently to conclusions about Faraday's natural theology, which are compatible with those presented here. (Clark, *Hibbert J.*, 144–7 (1967), and *Christian Graduate* 26–7 (1967).)
[42] This term is intended to embrace all theories identifying matter with force—including point atomism.

may be regarded as mechanical, or, equally, as electrical, since affinity is merely one aspect of a multifaceted unity, the 'powers of matter'.

Matter and the theology of nature

Faraday was not given to the study of speculative philosophy. His *Journal* and *Diary* are empty of metaphysics. Enthusiastic exaggeration has even acclaimed his scientific method as 'destined to purify science from the last remnant of metaphysics'.[43] Faraday, however, did not identify religion and natural theology with metaphysics. He was of the Sandemanian faith, the adherents of which followed the evangelical teachings of John Glas and Robert Sandeman; the Gnostic undertones (implicit even at the beginning of the nineteenth century) of *Naturphilosophie* were incompatible with the beliefs professed by Sandemanian Christians,[44] and, if Faraday had been philosophically at all receptive and sophisticated, he would have been aware of this. Although his conclusions often coincided with those of the Nature Philosophers, it is surely improbable that he would have been interested in their ideas as a justification of his theories about the physical world. Yet he held these theories with almost unreasoning tenacity.

Part of the explanation may be that a dynamical, unified theory fitted in with the world picture imposed by his religion. Faraday was deeply religious[45]—perhaps even more so than Davy.[46] I have already suggested that, for Davy, the power of God was the ultimate cause of the powers of matter, and that the connection between these powers was intimate. God's activity in the universe was not restricted to an initial act of creation, but was rather a constant presence. Just such an outlook had led Davy's contemporary Thomas Exley to remark that if matter existed solely by its powers (he advocated his own brand of force atomism), created by God, then the existence of matter was unremittingly maintained by the power of God.[47]

Religious convictions can and do play a major role in the selection of scientific theories, even though they do not further affect their content. A good example is provided by Joseph Priestley, whose theology

[43] H. Helmholtz, *J. chem Soc.* **34**, 277 (1881).
[44] —or indeed with any Christian beliefs! For a good account of the Sandemanian sect, see L. P. Williams, *Michael Faraday*, pp. 2–6 (Chapman and Hall, London, 1965).
[45] ibid., pp. 102–6.
[46] While on tour with Davy, Faraday wrote to Benjamin Abbott from Geneva (6 Sept. 1814: quoted by Bence-Jones, op. cit. (3), vol. 1, p. 157): 'Travelling . . . I find, is almost inconsistent with religion (I mean modern travelling), and I am yet so old-fashioned as to remember strongly (I hope perfectly) my youthful education . . .' Davy was not so troubled.
[47] T. Exley, *Principles of Natural Philosophy*, p. xxvii. London (1829). A good account of Exley's dynamical theories is to be found in D. M. Knight, *Atoms and Elements*, pp. 65–70. London (1967).

encouraged his acceptance of a form of Boscovichean atomism.[48] Boscovich himself was well aware of the significance of his theory for natural theology, as is clear from his complaints against Priestley for theological misuse of his ideas,[49] and from the appendix to his *Theoria*, 'Relating to Metaphysics: the Mind and God'.[50] The influence of Faraday's religious convictions on his life and thought—including scientific thought—cannot be ignored. Indeed, Stewart and Tait cited Newton and Faraday together as splendid examples of the compatibility of science with religion.[51] God created and sustained the universe; a single Architect implied a unified and purposeful plan. God worked through powers and through forces, and surely these agreed much better with a universe of force atoms than with one of billiard-ball atoms. Such an association of the power of God with natural forces had been a commonplace interpretation to British 'natural-philosopher–theologians' from the time of Newton until the end of the eighteenth century,[52] and throughout Faraday's writings there are passages suggesting that his ideas were firmly in this tradition.[53] Perhaps the clearest statement occurs at the end of his Easter course of lectures at the Royal Institution in 1847:

In conclusion, I may remark that, whilst considering the state and condition with which matter is endowed, we cannot shut out from our thoughts the consequences as far as they are manifested to us, for we find them to be always for our good; neither ought we to do so, for that would be to make philosophy barren as to its true fruits. And when we . . . perceive that all this is done by virtue of powers in the molecules which are indestructible, and by laws of action the most simple and unchangeable, we may well, if I may say it without irreverence, join *awe and trembling* with *joy and gladness*.

Our philosophy, feeble as it is, gives us to see in every particle of matter, a *center* of force reaching to an infinite distance; binding worlds and suns together, and unchangeable in its permanency. Around this same particle we see grouped the powers of the various phenomena of nature: . . . the harmonious working of all these forces in nature, until at last the molecule rises up in accordance with the mighty purpose ordained for it, and plays its part in the gift of *life itself*. And therefore our philosophy, whilst it shows us these things, should lead us to think of Him who hath wrought them; for it is said

[48] See p. 15 above.

[49] R. E. Schofield, *A Scientific Autobiography of Joseph Priestley, 1733–1804*, pp. 166–71, M.I.T. (1966).

[50] R. J. Boscovich, *Theory of Natural Philosophy*, pp. 187–96. M.I.T. (1966).

[51] D. Stewart and P. G. Tait, *The Unseen Universe*, preface (1876).

[52] For a good account of this see H. Metzger, *Attraction universelle et religion naturelle chez quelques commentateurs anglais de Newton.* . . . Paris (1937).

[53] e.g. *Experimental Researches in Electricity.* 3 vols., London (1839–55). Sections 2447 and 2968 imply the acceptance of an economical theological teleology. Also, *Lectures on the Non-Metallic Elements*, ed. Scoffern, p. 2 (London, 1853), tells us that chemistry is admirably suited to awaken within us 'the sentiment of immortality'.

by an authority far above even that which these works present, that *the invisible things of Him from the creation of the world are clearly seen, being understood by the things that are made, even His eternal power and Godhead.*[54]

This passage, in the language of natural theology,[55] is the culmination of a lecture on the appositeness of the chemical affinities of the different elements for their purposes on earth, an argument also appearing in Paley's classic, *Natural Theology.*[56]

The erroneous opinion that Faraday was well able to separate science from religion seems to be widely held,[57] though expressing at best a partial truth and stemming apparently from Faraday's statement that

there is no philosophy in my religion. I am of a very small and despised sect of Christians, known if known at all, as Sandemanians, and our hope is founded on the faith that is in Christ. But though the natural works of God can never by any possibility come in contradiction with the higher things that belong to our future existence, and must with everything concerning Him ever glorify Him, still I do not think it at all necessary to tie the study of natural science and religion together, and in my intercourse with fellow-creatures, that which is religious, and that which is philosophical, have ever been two distinct things.[58]

This statement of his views on the relation of religion to science is strikingly similar to that made by Sir William Rowan Hamilton in a letter of about 1830.[59] Hamilton, however, did indeed conceive of a connection between natural theology and natural science. In 1832 he had written to Coleridge that *'objectively, and in the truth of things, the powers attributed to atoms belong not to them but to God'* ;[60] he was both a Christian and an Idealist, and was inclined to Boscovichean atomism

[54] R. Instn MS. 23; quoted by Bence-Jones, op. cit. (3), vol. 2, pp. 229–30.

[55] cf. R. E. D. Clark, *Hibbert J*. 147 (1967). But see note (69) below.

[56] W. Paley, *Natural Theology*, pp. 69–73. 1st edn, not annotated (1802), this edn, with notes by H. Brougham and C. Bell, 2 vols., London (1836). Perhaps the finest relation of chemical affinity to the argument from design occurs in S. Parkes *Chemical Catechism*, p. 432 n. (10th edn, London, 1822):

The varied influence of this property of matter [chemical affinity] may be attributed by the atheist to *chance*; but the man of sober reflection, who allows the evidence of a mass of facts to have its natural influence upon his mind, will be persuaded that chemical affinity can neither be ascribed to accident, nor to a necessity in the nature of things; for he will attribute the whole to the contrivance, to the wisdom, and to the goodness of an intelligent Agent, who has varied these operations in a thousand ways, to suit the designs of his beneficence, and to promote different and distinct purposes of utility and happiness.

Parkes was a great admirer of Davy and of his theories.

[57] e.g. C. C. Gillispie, *Genesis and Geology*, p. 208. Harper Torchbook (1959).

[58] Quoted by Bence-Jones, op. cit. (3), vol 2, pp. 194–5.

[59] See note (63).

[60] Quoted by R. P. Graves, *Life of Sir W. R. Hamilton*, vol. 1, p. 593 (my italics). 3 vols., London and Dublin (1882–9).

because of its religious associations, and also because Boscovich's views
'seem capable of being incorporated with high metaphysical idealism'.[61]
In the year when he declared this, 1834, Hamilton first met Faraday,
and discovered with delight that they had almost identical views of the
nature of matter.[62]

To return to the theological argument: Hamilton's concern for the
distinction between religion and philosophy was merely to ensure that

no reasonable complaint lies against religion for not presenting articles of
faith under the form of Philosophical Theorems, though we may justly
expect that the doctrines which it teaches should not be contradictory, nor
capable of being proved to be absurd:—an *a priori* requisition of the mind
with which it is my intellectual belief that Christianity complies.[63]

This necessary postulate of non-contradiction was the active agent of
selection operated by religion upon science; it was, moreover, as we
have seen, a postulate upon which Faraday was equally insistent, and
the negation of which would have seemed to him to be an utter impossibility. Had not Robert Sandeman held that Christianity was of an
intellectual character, having its origins in the understanding?[64] The
conclusion drawn by Sandeman, and probably by all evangelicals,[65]
was that the validity of natural theology was to be endorsed—the
natural world must display an intellectually comprehensible unity and
logical coherence. One could interpret the natural world in terms of the
primary facts of God's existence and of His revelation to man. This
argument follows precisely the opposite direction to that of Paley, who
thought that one could induce some of the divine attributes from an
examination of the order and purpose displayed in the physical world.[66]
Faraday explicitly repudiated this conventional version of natural
theology,[67] and preferred to argue from anterior knowledge of God to
deductions about nature.[68] He adopted the theology of nature, rather

[61] Quoted by R. P. Graves, op. cit. (60), vol. 2, p. 86: letter of 27 June 1834.
[62] ibid., vol. 2, pp. 95–6: letter of 30 June 1834.
[63] ibid., vol. 2, pp. 397–8: letter of 1835–6.
[64] J. Stoughton, *Worthies of Science*, p. 273. Religious Tract Society, London.
[65] Clark, op. cit. (55), 145. But see note (69) below.
[66] e.g. Paley (ed. cit. (56), vol. 2, p. 101): 'Of the "Unity of the Deity", the proof is,
the uniformity of plan observable in the universe.'
[67] Faraday, at the commencement of a lecture on Mental Education in 1854:

Let no one suppose for a moment that the self-education I am about to commend, in
respect of the things of this life, extends to any considerations of the hope set before us,
as if man by reasoning could find out God. It would be improper here to enter upon this
subject further than to claim an absolute distinction between religious and ordinary
belief. I shall be reproached with the weakness of refusing to apply those mental operations which I think good in respect of high things to the very highest. I am content to
bear the reproach.

(Quoted by Silvanus P. Thompson, *Michael Faraday*, p. 292. London (1898).)
[68] cf. Williams, op. cit. (44), p. 103.

than natural theology.[69] But, in either case, the thorough-going divorce of science from religion made absolutely no sense, nor was it attempted by Faraday, who realized that to distinguish science from religion was not to sever them, but only to indicate the latter's absolute and logical primacy, while limiting the former's sphere. In spite of Faraday's protestations, his deepest intuitions about the physical world arose from his faith in the 'Divine origin of nature'.[70]

Faraday was certainly too honest and methodical a natural philosopher to allow external factors, be they metaphysical or religious, to trespass overtly into his laboratory. It may, however, be significant that, in a personal memoir on the nature of matter,[71] he thrice alludes to God in support of the possibility of point atoms. 'Is the lingering notion which remains in the minds of some, really a thought, that God could not just as easily by his word speak power into existence around centers, as he could first create nuclei and then clothe them with power?'

Electrochemical affinity

We are now almost ready to consider Faraday's greatest scientific work, embodied in his *Experimental Researches in Electricity*. The years prior to 1831 were Faraday's years of study and preparation, fashioning the basis, in fact and hypothesis, for his best laboratory work. He regarded chemical affinity as an important facet of the powers of matter; his theory of matter, grown more dynamical—though not yet purely so —lay within a framework of Newtonian natural theology. The debt to Davy is clear, as also is the imprecise delineation of his dynamical theory of matter. The latter requires some development here.

First, Faraday, although very much Davy's pupil, was beginning to develop his own critical approach, even towards his master's ideas. He was 'naturally sceptical on philosophical theories',[72] and in general required firm experimental proof before he, as a 'cautious philosopher', felt justified in receiving any theory into his stock of knowledge.[73] He made these comments about Ampère's electrodynamical theory of electricity, magnetism, and chemical affinity,[74] but they might equally well have been applied, at this date, to Davy's flirtation with Boscovichean atomism.

[69] The evangelical attitude towards natural theology was that it was secondary in importance to revelation: unless subjected to rectification through the spectacles of revealed religion (Calvin's term), natural theology was valueless. In contrast, God's revelation to man could directly yield information about the natural world. Thus, whereas the theology of nature was valid without qualification, this was not true of natural theology.   [70] cf. Williams, op. cit. (44), p. 4.

[71] Instn elect. Engrs. MS. dated Feb. 1844. Published *Br. J. Hist. Sci.* **4**, 105–7 (1968).

[72] Faraday to de la Rive, 1821: quoted by Bence-Jones, op. cit. (3), vol. 1, pp. 354–7.

[73] Faraday to A. Marcet, 1822: R. Instn MS. 11.   [74] See Chapter 4.

A M—G

Faraday's scepticism sometimes reached even his own conclusions. He and Wollaston, two of the most cautious scientists of their time, who by their work, may have helped to weaken chemists' belief in 'Daltonian' atoms,[75] both furnished 'proofs' that matter was atomic. Wollaston had argued that the atmosphere, being finite in height, must be made up of finite particles; in other words, matter was atomic.[76] Faraday argued that if matter was corpuscular, then there should be a 'limit of vaporization' below which the attractions of gravity and cohesion would exceed the repulsion of heat. He found that there was such a limit; below a certain temperature, any given body exerted no vapour pressure. He concluded that matter must be discontinuous.[77] (Both these arguments are based on the fallacy that an infinite series cannot have a finite sum; this was one of the few occasions when Faraday was seriously hampered by his lack of mathematical training.)[78] However, he remained cautious: 'I refrain from extending these views, as might easily be done, to the atomic theory, being rather desirous that they should first obtain the sanction or correction of scientific men.'[79] Even this mild apostasy from dynamism did not please Davy;[80] Herschel, however, and, predictably, Wollaston, were quite satisfied.[81] In the later years of the decade of the 1820s, however, Faraday gradually reverted to a more continuous dynamical–mechanical picture of chemical affinity.[82]

On 29 November 1831 Faraday wrote to his friend Phillips from Brighton,

We are here to refresh. I have been working and writing a paper that always knocks me up in health,[83] but now I feel well again, and able to pursue my

[75] cf. W. Brock, ed. *The Atomic Debates*. Leicester (1967); D. M. Knight, *Atoms and Elements*. London (1967).

[76] M. Faraday, *Phil. Trans. R. Soc.* **112**, 89–98 (1822).

[77] M. Faraday, *Phil. Trans. R. Soc.* **116**, 484–93 (1826).

[78] Faraday seldom seemed to regard his lack of mathematical training as a drawback; only when writing to Clerk Maxwell, and when discussing Ampère's theory, does he seem to feel hampered. In the early 1830s he wrote to Mary Somerville (Bodleian Library, Somerville papers VI): 'I do not remember that Mathematics have *predicted* much. Perhaps in Ampère's theory one or at most 2 independent facts. I am doubtful of 2. Facts have preceded the mathematics or where they have not the facts have remained unsuspected though the calculations were ready.'　　　[79] M. Faraday, op. cit. (77).

[80] R. Soc. MS. vii, 169: Faraday to J. F. W. Herschel, 26 May 1826.

[81] ibid.: Herschel to Faraday, 26 May 1826, and (80).

[82] In 1826 he wrote to Daubeny that 'pressure would rather produce combination than destroy it. Indeed I have two or three facts or rather Mr. Brande has confirmation of the latter opinion.' (2 May 1826: R. Instn MS. 25.) In the following year he suggested that sudden expansion of gaseous hydrocarbons led to partial decomposition. 'If this explanation should ultimately prove, by further experiments, to be true, it will be highly important, as affording an instance of the exertion of mechanical and chemical powers in those circumstances where they most closely verge upon each other.' (*Q. J.Sci.* NS. **1**, 204–6 (1827).)　　　[83] *Phil. Trans. R. Soc.* **122**, 125–62 (1832).

subject; and now I will tell you what it is about. The title will be, I think, 'Experimental Researches in Electricity':—

I. On the Induction of Electric Currents;
II. On the Evolution of Electricity from Magnetism;
III. On a new Electrical Condition of Matter;
IV. On Arago's Magnetic Phenomena. There is a bill of fare for you; and, what is more, I hope it will not disappoint you.[84]

Faraday wound two insulated lengths of copper wire around a block of wood, attaching the ends of one to a galvanometer and of the other to a battery. A momentary current was induced in the secondary helix on making or breaking the primary circuit.[85] Analogous effects were observed when a magnet was introduced into or removed from a wire helix connected to a galvanometer.[86] Whilst the wire was subject to either volta-electric or magneto-electric induction there obtained '*a peculiar state*'.[87] This peculiar state was undetectable;[88] yet Faraday, although always sceptical about philosophical theories, clung to the hypothesis of such a state's existence. Without it, he would have been forced to accept fluid theories of electricity; by now, however, he had sufficient confidence in dynamism to reject imponderable fluids.[89]

The new electrical condition which intervenes by induction between the beginning and the end of the inducing currents gives rise to some very curious results. It explains why chemical action or other results of electricity have never been as yet obtained in trials with the magnet. In fact, the currents have no sensible duration. I believe it will explain perfectly the *transference of elements* between the poles of the pile in decomposition . . . The condition of matter I have dignified by the term *Electronic*, THE ELECTRONIC STATE. What do you think of that?[90]

Faraday described the electrotonic state as a state of strain in the *particles*,[91] and suggested that electric impulses might arise from 'the momentary propulsive force exerted by the particles during their arrangement'.[92] An electric current might thus be regarded as a succession of arrangements and rearrangements of molecular forces within the conductor. This hint was easily developed to a hypothetical explanation of electrolytic transfer. Suppose that when a current was

[84] Bence-Jones, *op. cit.* (3), vol. 2, pp. 6–10.
[85] M. Faraday, *Experimental Researches in Electricity*, para. 10. 3 vols., London (1839–55). Faraday called this 'volta-electric induction'. For a detailed account of the discovery of electromagnetic induction, see Williams, op. cit. (44), chapters 4 and 5.
[86] Faraday, op. cit. (85), para. 38. Faraday called this 'magneto-electric induction'.
[87] Bence-Jones, op. cit. (3), vol. 2, pp. 6–10.   [88] Faraday, op. cit. (85), para. 61.
[89] Williams, op. cit. (44), pp. 198 ff.   [90] Bence-Jones, op. cit. (3), vol. 2, pp. 6–10.
[91] Faraday, op. cit. (85), para. 73.   [92] ibid., para. 73.

passed through electrolytes the latter were thrown into the electrotonic state.[93] If the electrolyte was decomposable, the resulting strain might be sufficient

to make an elementary particle leave its companion, with which it is in a constrained condition, and associate with the neighbouring similar particle, in relation to which it is in a more natural condition, the forced electrical arrangement being itself discharged or relieved, at the same time, as effectually as if it had been freed from induction. But as the original voltaic current is continued, the electro-tonic state may be instantly renewed, producing the forced arrangement of the compound particles, to be instantly discharged by a transference of the elementary particles of the opposite kind in opposite directions, but parallel to the current. . . . But as I have reserved this branch of the enquiry, that I might follow out the investigations contained in the present paper, I refrain (though much tempted) from offering further speculations.[94]

Small wonder that Faraday was tempted to further speculations! If his hypothesis proved correct, then he would have a complete theory of electrolysis, incidentally explaining electricity and chemical affinity as different aspects of the powers of matter; as an additional bonus, powerful support would accrue for a dynamical theory of matter, and the unity of nature would be more fully revealed! The speculative content of his first series of *Experimental Researches in Electricity* was even more exciting than the solid content of experimental proof.[95]

Faraday alternately rejected and returned to his electrotonic state, while remaining constant in conceptual approach. Fundamental to the concept of this electrotonic state, as to the theories of chemical and electrical forces following from it, were the concepts of mechanical arrangement of particles and powers—concepts that, within a dynamical theory of matter, became identical with one another. In his second series of researches, read in January 1832, he abandoned his early hypothesis of an internal state of strain for a field hypothesis, where the strain was transferred to lines of magnetic and electrical force.[96] This transfer of emphasis (it was not really a change of attitude) in no way weakened his conviction that there had to be an

Influence of Mechanical arrangement over chemical affinity. Is not introduction of wire or nucleus, solid, into saturated cold solutions or cooled acetic acid the same principle more highly exalted?
    And then is not that the same as action of spongy platina on oxygen and

[93] This is anticipating Faraday's later nomenclature; but since this was intended to be theory-free, no confusion should result.
    [94] ibid., para. 76.
    [95] cf. Williams, op. cit. (44), p. 200.
    [96] Faraday, op. cit. (85), para. 231.

hydrogen. If either gas alone were put up into spongy platina would there not be condensation?

Are not all these phenomena results of one cause?

Make crystals of salts with wires, etc. in them. Put spongy platina into gases which are required to combine and subject all to sunshine.[97]

Faraday's explanation of catalysis was mechanical in a special way that was all his own;[98] it might equally well have been termed electrical, for he understood mechanical powers within the embracing framework of the powers of matter. Chemical, mechanical, and electrical forces were ultimately the same; this was his conviction, and it was his self-imposed task to reveal their identity.

Faraday's third series of researches took him a little further along the road to this demonstration.[99] He proved that electricity was the same regardless of its origin; voltaic, static, magneto-, thermo-, and animal electricity really differed only in quantity and in intensity, and these factors could account for their apparently different effects.[100] One of the tests that he used for demonstrating the identity of electricities was chemical decomposition,[101] effected by 'electricity in motion'. While implicitly relating electrical and chemical powers to one another, he defined the electric current as 'anything progressive, whether it be a fluid . . . or merely vibrations, or, speaking still more generally, progressive forces. By *arrangement* I understand a local adjustment of particles, or fluids, or forces, not progressive.'[102] Quantitative experiments suggested that, probably for all cases of electrochemical decomposition, *'the chemical power . . . is in direct proportion to the absolute quantity of electricity* which passes'.[103] This was a valuable correlation in Faraday's search for underlying unity; chemical affinity and electrical force were thereby brought closer together. He moved eagerly and immediately to a closer correlation of chemical, electrical, and mechanical power.

In December 1832 he asked in his *Diary*:

218. Can an electric current, voltaic or not, decompose a solid body, ice, etc., etc.

219. If it can, does it give structure at the time.

[97] Faraday, op. cit. (38), vol. 1, pp. 360–1, entry of 1 Aug. 1832.

[98] See also notes (159)–(162) below.

[99] *Phil. Trans. R. Soc.* **123**, 23–54 (1833).

[100] The distinction between quantity and intensity of electricity was first recognized by Cavendish (*Phil. Trans. R. Soc.* **51**, 584 (1771), and **66**, 196 (1776)), but thereafter frequently ignored. Berzelius, for example, as Sir Harold Hartley has remarked (British Association Centenary Meeting, London 1931: presidential address to chemical section, p. 13), failed to understand Faraday's laws of electrolysis because he did not understand this distinction.

[101] Faraday, op. cit. (85), para. 267.

[102] ibid., para. 283.   [103] ibid., para. 377.

His experiments soon led him to the generalization that substances that conducted as fluids became insulators on solidification. Thus, for example, solid silver chloride was an insulator, fused silver chloride a conductor.[104] The general assumption of conducting powers on lique-faction 'seems importantly connected with some properties and rela-tions of the particles of matter'. Faraday suspected that the pheno-menon arose because particles in the solid state, lacking freedom of movement, would not undergo the rearrangements requisite for electro-lysis.[105] In his *Diary* he went further, and speculated about the general significance of his results:

15 FEBY. [1833]
285. The difference between solid and fluid state shews most important rela-tion and opposition between mechanical aggregation and chemical affinity, i.e. of that affinity modified by voltaic power. First may in some respects be considered even *more powerful* than the latter. Again, is the reverse as to heat, where fluid *does not* conduct and solid *does*.
286. Does not insulation by solid shew that decomposition by V. pile is due to slight power super added upon previous chemical attractive forces of particles when fluid? Since mere fixation of particles prevents it, it must be slight.
287. Does it not shew very important relation between the decomposability of such bodies and their conducting power, as if here the electricity were only a transfer of a series of alternations and vibrations and *not* a body transmitted directly. May settle or relate to question of materiality or fluid of Electricity.

Faraday, believing that electrochemical decomposition might arise from the addition of electrical power to existing chemical powers, sought an explanation that would embrace also his picture of electrolysis as effected by a series of compositions and decompositions. He had be-come sure that the latter was a true interpretation of the phenomena in October 1832, when he carried out electrolyses in a jelly, examining the progress of electrolytic transfer by means of indicators. He found that the substances were not transferred instantly and in a free state, and concluded (it would seem on insufficient evidence) that the process must therefore have occurred 'gradually and by successive combination and decomp'.[106]

This, together with the conclusion drawn from his fourth series, pre-pared him for the announcement of his theory of electrolysis. He introduced his paper[107] 'On Electrochemical Decomposition' by reject-ing the generally held theory that poles of attraction and repulsion were

[104] Faraday, op. cit. (85), para. 402. cf. H. Davy, *J. R. Instn*, 53 (1802).
[105] ibid., paras. 412–13.
[106] Faraday, op. cit. (38), entry of 26 Oct. 1832.
[107] *Phil. Trans. R. Soc.* 675–710 (1833).

essential to electrolysis. His experiments dating from September 1832 proved that poles were not necessary to electrolysis.[108]

Hence it would seem that it is not a mere repulsion of the alkali and attraction of the acid by the positive pole, etc. etc. etc., but that as the current of electricity passes, whether by metallic poles or not, the elementary particles arrange themselves and that the alkali goes on as far as it can with the current in one direction and the acid in the other.[109]

The results suggested that there was an internal action of the parts suffering decomposition, and 'that the power which is effectual in separating the elements is exerted there, and not at the poles'.[110]

Before announcing his own theory of electrochemical decomposition, he reviewed the principal alternatives. Grotthus had considered the pile as an electric magnet, whose poles exerted attractive and repulsive forces on the particles in solution.[111] Since poles were unnecessary for electrolysis, this theory could not be true. Davy's theory seemed to refer to action arising from poles, but transmitted from particle to particle.[112] Faraday, rejecting poles, objected also to the extreme generality of this theory, necessitated by the state of knowledge in 1806, but no longer justifiable. A. de la Rive considered that electrolysis occurred as a result of the interplay of ordinary chemical affinities between the particles of the electrolyte and of the two material electricities passing through the solution.[113] Any theory requiring imponderable fluids was bound to incur Faraday's disapproval.

Thus Faraday rejected his predecessors' theories of electrolysis. They were unsatisfactory, relying on action from poles, now experimentally discredited, or including unjustified assumptions about the nature of electricity. Faraday therefore sought a new and better theory. Electrolysis was effected by the electric current, whose definition would be a first step. 'It . . . may perhaps best be conceived of as *an axis of power having contrary forces, exactly equal in amount, in contrary directions.*'[114] We have already encountered the various speculations and experiments responsible for the elements of Faraday's theory of electricity; he now proceded to their synthesis.

Since chemical substances were held together by chemical affinity, which in electrolysis was overcome by the electric current, and, furthermore, since electrolysis apparently resulted from action at the particles, Faraday concluded that the effect was produced by 'an *internal corpuscular action,* exerted according to the direction of the electric current', and that it was due to a 'force either *superadded to, or giving*

[108] Faraday, op. cit. (38), paras. 81 ff., especially 104.
[109] ibid., para. 103.   [110] Faraday, op. cit. (85), para. 471.
[111] *Ann. Chim. Phys.* **58,** 64 (1806).   [112] See Chapter 2.
[113] *Ann. Chim. Phys.* N.S. **28,** 190 (1825). [114] Faraday, op. cit. (85), para. 517.

direction to the ordinary chemical affinity of the bodies present'.[115] 'In this view the effect is considered as *essentially dependent* upon the *mutual chemical affinity* of the particles of opposite kinds.'[116] Decomposition occurred along straight or curved[117] lines followed by the current between the poles. 'The theory which I have ventured to put forth (almost) requires an admission, that in a compound body capable of electrochemical decomposition the elementary particles have a mutual relation to, and influence upon each other, extending beyond those with which they are immediately combined.'[118] In spite of his habitual caution, he had brought himself to the edge of revealing explicitly the dynamical basis of his theory of matter. He still, however, gave no clear indication of the precise brand of dynamism involved.

I hope I have now distinctly stated, although in general terms, the view I entertain of the cause of electro-chemical decomposition, *as far as that can at present be traced and understood.* I conceive the effects to arise from forces which are *internal*, relative to the matter under decomposition—and not *external*, as they might be considered, if directly dependent upon the poles. I suppose that the effects are due to a modification, by the electric current, of the chemical affinity of the particles through or by which that current is passing, giving them the power of acting more forcibly in one direction than in another, and consequently making them travel by a series of successive decompositions and recompositions in opposite directions, and finally causing their expulsion or exclusion at the boundaries of the body under decomposition, *and that* in larger or smaller quantities, according as the current is more or less powerful . . .

Having thus given my theory of the mode in which electro-chemical decomposition is effected, I will refrain for the present from entering upon the numerous general considerations which it suggests, wishing first to submit it to the test of publication and discussion.[119]

The temptation to give full scope to his imaginative and speculative faculties was strong; so was Faraday's self-discipline. 'I must keep my researches really *Experimental*', he wrote that winter, 'and not let them deserve any where the character of *hypothetical imaginations.*'[120] The hypothetical imaginations would probably have ranged over the topics of chemical affinity and electricity, and their interrelation. Meanwhile, there was the laboratory problem of bringing respectability to esoteric hypotheses about matter and force.

Faraday was clearly attracted by the 'beautiful theory'[121] put forward by Davy and by Berzelius that ordinary chemical affinity 'is a mere result of the electrical attractions of the particles of matter',[122]

[115] Faraday, op. cit. (85) para. 518.　　[116] ibid., para. 519.
[117] ibid., paras. 521–2.　　[118] ibid., para. 523.
[119] ibid., para. 524.　　[120] Faraday, op. cit. (38), para. 1207.
[121] Faraday, op. cit. (85), para. 703.　　[122] See Chapter 5.

and was troubled by the refusal of some substances to submit to electrolytic decomposition. If they would not decompose, it might be because they could not—perhaps because they were non-conductors.[123] Faraday wanted to explain this apparently anomalous behaviour—if electricity and affinity were manifestations of the same power, why did the former sometimes fail to modify the latter, why could not these two aspects of a single power of matter always react upon one another?

Although troubled by such problems, he continued to search for positive evidence in support of a more precise picture of electrolysis, and a tighter correlation between chemical and electrical force. He sought for, and found, additional confirmation of his first law of electrolysis. He produced what he regarded as 'an irresistible mass of evidence, proving the truth of the important proposition . . . *that the chemical power of a current of electricity is in direct proportion to the absolute quantity of electricity which passes*'.[124] Since the results held generally for all electrolytes, the latter could be arranged into a single series, where the individual substances occupied definite positions, corresponding to the precise degree of their chemical affinities.

Faraday now called the 'particles' formed by electrolytic dissociation *ions*. 'They are combining bodies; are directly associated with the fundamental parts of the doctrine of chemical affinity; and have each a definite proportion, in which they are always evolved during electrolytic action.'[125] The numbers representing the proportions in which ions were evolved in electrolysis were termed *electrochemical equivalents*, which 'coincide, and are the same, with ordinary chemical equivalents'. The significance of this relation was not lost upon Faraday. Excitedly he wrote:

I think I cannot deceive myself in considering the doctrine of definite electro-chemical action as of the utmost importance. It touches by its facts more directly and closely than any former fact, or set of facts, have done, upon the beautiful idea, that ordinary chemical affinity is a mere consequence of the electrical attractions of the particles of different kinds of matter.[126]

He noted also the desirability of tabulating the electrochemical relations of bodies, because of its importance 'as to nature of chemical affinity and its relations to electrical states and powers'.[127]

A further deduction from his quantitative results was that the individual particles of matter were associated with a determinate quantity of electricity; the probable identity of chemical and electrical

[123] Faraday, op. cit. (85), 4th series, and the electrical relations between diamond and graphite. cf. note (165) below.   [124] ibid., para. 821.
[125] ibid., para. 824.   [126] ibid., para. 850.
[127] Faraday, op. cit. (38), para. 801.

powers, demonstrated by his own researches, coupled with what Faraday called 'the teaching of Dalton, that chemical powers are . . . definite for each body', made such a conclusion almost inevitable. Had Faraday believed in Daltonian atoms, he would presumably have accepted electrical fluids, and would have deduced that electricity was particulate—he would have deduced the existence of the electron. However, his dynamical preconceptions prevented him from saying more than that 'the atoms of matter are in some way associated with electrical powers, to which they owe . . . their mutual chemical affinity'.[128] The correlation between definite proportions and definite electrochemical action introduced great

harmony . . . into the associated theories of definite proportions and electrochemical affinity . . . According to it, the equivalent weights of bodies are simply those quantities of them which contain equal quantities of electricity, or have naturally equal electric powers; it being the ELECTRICITY which *determines* the equivalent number, *because* it determines the combining force. Or, if we adopt the atomic theory or phraseology, then the atoms of bodies which are equivalents to each other in their ordinary chemical action, have equal quantities of electricity naturally associated with them. But I must confess I am jealous of the term *atom*; for though it is very easy to talk of atoms, it is very difficult to form a clear idea of their nature, especially when compound bodies are under consideration.[129]

Faraday was indeed unsure of any atomic theory; he had thought long and closely about theories of attraction and of particles and atoms of matter, but, as he wrote to a young student of King's College, London, 'the more I think (in association with experiment) the less distinct does my idea of an atom or particle become'.[130] However, the very fact that he found it more difficult to picture compound bodies than simple ones indicates that he was by now thinking exclusively in dynamical terms; Daltonian atoms remained simple, however many of them were placed together, while force atoms in combination would interact to produce force patterns of a complexity that increased alarmingly as the molecules contained more atoms.

Faraday had been in constant correspondence with Whewell in this period; the latter had advised him on the new terminology he required to express his ideas on electrolysis.[131] Whewell was absolutely delighted at this seventh series of researches, with its statement of the doctrine of definite electrochemical action. He wrote to Faraday congratulating him on his researches:

[128] Faraday, op. cit. (85), para. 852.   [129] ibid., para. 869.
[130] Letter to F. O. Ward, 16 June 1834: cited by Hofmann in the Faraday Lecture for 1875.
[131] See Faraday–Whewell correspondence, Trinity College, Cambridge; and Williams, op. cit. (44), pp. 261–6.

I have all along considered them as the greatest event which ever happened in the history of chemistry. It has for some time been clear that the capital point of chemical theory is the connexion between electrical relation and chemical composition; and you have now got so far as to obtain numerical measure of the former relation; and some facts belonging to its connexion with the numerical laws of composition. It cannot be doubted therefore that you have made great steps towards the solution of the grand problem; and my own persuasion is that you have before you still greater discoveries.[132]

Whewell was something of a Kantian, to a degree then unfashionable in England; he was eager for the interrelation and unification of scientific phenomena and laws, and was convinced that the universal concept of *polarity* would provide the instrument for this unification.[133] Electrical and magnetic powers were recognized as polar; now Faraday had provided concrete experimental evidence relating them to chemical powers, and the grand concept of polarity could be extended to chemical phenomena. Whewell's enthusiasm was a little in advance of the discoveries he anticipated; but his instinct was, on this point at least, sure. Faraday really did have great discoveries ahead of him, and polarity was to assume greater significance in his work.

He had first to convert the beautiful probability of the identity of chemical and electrical powers into a certainty; correlation was not sufficient, identity must be demonstrated. To achieve this, it was first necessary to identify the source of electricity in the voltaic pile. The state of opinion when Faraday approached the problem was extraordinarily confused. The investigator would encounter, in the current literature, 'such contradictory evidence, such equilibrium of opinion, such variation and combination of theory, as would leave him in complete doubt respecting what he should accept as the true interpretation of nature'.[134]

Volta had proposed, in the letter announcing his discovery of the pile, that the electric current arose from the contact of dissimilar metals with the electrolyte;[135] Fabbroni differed utterly, and argued that the effect was chemical in origin.[136] Davy attempted to reconcile these opposing theories by showing that they were both partial truths, individually inadequate, but in combination providing the true explanation.[137] The issue was important, and controversy continued long after Davy's attempted synthesis. Brande[138] had seen the cause in the

---

[132] Letter of 25 Sept. 1835: Trinity College, Cambridge MS. holograph letter: R. Instn MS. Faraday's 'Portrait Book'.

[133] See Chapter 4 below, and Whewell's *Philosophy of the Inductive Science*, vol. 1, pp. 331–60. 2 vols., London (1840).   [134] Faraday, op. cit. (85).

[135] *Phil. Trans. R. Soc.* **90**, 403 (1800).   [136] *Nicholson's J.* **4**, 120 (1800).

[137] See Chapter 2.

[138] W. T. Brande, *A Manual of Chemistry*, vol. 1, p. 279. 3 vols., London (1821).

contact of metals, as had Biot,[139] Becquerel, and Ampère.[140] There were some, like Savary, who thought that the issue could be resolved once it had been ascertained whether current electricity consisted of a separate matter, or merely of vibrations.[141] Faraday's opinion, that there was no separate matter of electricity, and that electricity was merely a power of matter, decided his preference. Since chemical and electrical powers appeared to be identical, it might seem pedantic to argue that chemical powers were the cause of voltaic electricity, rather than the reverse; but such a conclusion was necessary for the internal consistency of his ideas. He had rejected electrical fluids, and had dismissed poles as factors in electrolysis—the *electrodes*, as he now called them, merely represented the limiting surfaces of electrolytes, and electrolysis proceeded by an exertion of power by the individual particles. The voltaic current therefore arose, not from the contact of electrodes with electrolyte, but by an exertion of the chemical powers of the latter.[142] Although the opposite conclusion would have been disastrous for his hypotheses, and would have required fundamental re-thinking of his basic assumptions, Faraday did not prejudge the issue. He knew and accepted that experimental philosophy 'is a great disturber of pre-formed theories'.[143]

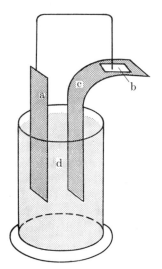

FIG. 3.1. (a) Platinum plate; (b) absorbent paper moistened with potassium iodide solution; (c) zinc plate; (d) dilute sulphuric acid.

His argument about the identity of chemical and electrical attraction disposed him to believe that the immediate supply of electricity was due to chemical powers; but the origin of the current was, for him, a different problem.[144]

How very needful the current is to the decomposition in a single pair of plates. How very needful for the existence of the current *is decomposition* . . . But as they cannot be cause and effect to each other, what is the common origin

[139] Biot, *Traité de physique expérimentale et mathématique*, vol. 2, pp. 478 ff. 4 vols., Paris (1816).
[140] *Ann. Chim. Phys.* **27**, 29 (1824). After Ampère had proposed his electrodynamical theory of electromagnetism (Chapter 4) his ideas became more complicated.
[141] *Ann. Chim. Phys.* **34**, 54 (1827).
[142] Faraday here overlooked the e.m.f. on open circuit, which *was* a contact phenomenon.
[143] See note (130) above.
[144] Faraday, op. cit. (85), para. 878.

of both? Must make this out. It is of no use continuing to suppose one as producing the other in either order.[145]

He first demonstrated that metallic contact was not necessary to electrolysis by moistening a piece of absorbent paper with potassium iodide solution, placing it on a bent zinc plate (see fig. 3.1), and touching it with a platinum wire attached to a platinum plate.[146] When the lower ends of both plates were dipped into a dilute acid solution, electrolytic decomposition of the iodide occurred; decomposition was polar, and dependent upon the direction of the current of electricity passing from the zinc through the acid to the platinum. That the electrolysing current was due to 'the state of things in the vessel', and to no other cause, was proved by (i) removing the metallic plates from the acid, when electrolysis ceased, and (ii) connecting the metallic plates directly together, when electrolysis of the potassium iodide again occurred, but in the reverse direction. These experiments suggested to Faraday 'a most extraordinary mutual relation of the chemical affinities of the fluid which *excites* the current, and the fluid which is *decomposed* by it'. He generalized this relation, and stated that he regarded electrolytic decomposition 'as being the direct consequence of the superior exertion at some other spot of the same kind of power as that to be overcome, and therefore as the result of an antagonism of forces of the same nature'.[147]

Now the only kind of power capable of producing chemical change was, by definition, chemical affinity. Since the generation of electric currents in batteries and electrolytic decomposition both resulted from the preponderance of one set of chemical affinities over another less powerful set, there was no need, when discussing the phenomena, to refer to any other power.

All the facts show us that that power commonly called chemical affinity, can be communicated to a distance through the metals and certain forms of carbon; that the electrical current is only another form of the forces of chemical affinity; that its power is in proportion to the chemical affinities producing it; that when it is deficient in force it may be helped by calling in chemical aid, the want in the former being made up by an equivalent of the latter; that, in other words, *the forces termed chemical affinity and electricity are one and the same*.[148]

'All the facts' included experiments with electrolytic cells, with additional plates inserted between the outside electrodes.

[145] Faraday, op. cit. (38), para. 1528.
[146] Faraday, op. cit. (85), paras. 880–7.
[147] ibid., para. 910.
[148] ibid., para. 918.

This arrangement (see fig. 3.2) retarded the current, seeming to show, as early as February 1834, the

antagonism of the *chemical powers* at the Electromotive parts with the *chemical powers* at the interposed parts. The first are producing electric effects; the second opposing electric effects, and the two seem equipoised as in a balance, and in both cause and effect appear to be identical with each other. *Hence chemical action merely electrical action and electric action merely chemical.*

. . . Chem. affinity and Electricity the same.[149]

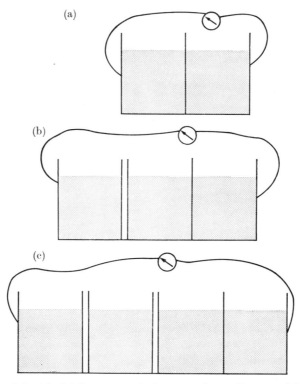

Fɪɢ. 3.2. (a), (b) Current entirely stopped; (c) Current able to pass.

Harking back to his electrotonic state, and to his notion of electrolysis as the arrangement and rearrangement of powers, Faraday forcefully suggested that, in an open voltaic cell, a single electrode in water set up an electrical double layer, with the particles in a peculiar state of tension or polarity.[150] Completion of the circuit would relieve the strain, whereupon a series of compositions and recompositions would follow— electrolysis would occur. Faraday, always mindful of the arrangement

[149] Faraday, op. cit. (38), paras. 1534, 1642.   [150] Faraday, op. cit. (85), para. 949.

of powers, suspected that the state of strain around a single electrode on open circuit might produce something like structure, and straightaway set about searching for such an effect, by using polarized light.[151] Failing to reveal any evidence of structure by this method, he concluded that there was 'no reason to expect that any kind of structure of tension can be rendered evident'[152]—but this was no reason for giving up the concept of such a state! Indeed, the cumulation of his experimental results necessitated a state of strain. Had Faraday started with different prejudices, without a unified dynamical background, he would scarcely have arrived at the idea of an electrical double layer in a state of polar strain.

The structure and powers of matter

One consequence of Faraday's continuous concern with structure and arrangement was his great interest in the 'Influence of Mechanical arrangement over chemical affinity'.[153] Clearly, in Faraday's scheme, mechanical arrangement and mechanical powers were intimately related—we have already encountered one occasion where he regarded mechanical and chemical powers as verging closely upon one another. Faraday sought the explanation of catalysis in this intermediate territory. Döbereiner had discovered the catalytic effect of platinum on a mixture of hydrogen and oxygen, explaining it as an electrical phenomenon.[154] This was generally accepted. Yet when Faraday announced his sixth series of experimental researches, 'On the power of Metals and other Solids to induce the Combination of Gaseous Bodies',[155] he did so with the reservation that 'remarkable as the phenomena are, the power which produces them is not to be considered as of an electric origin, *otherwise than as all attractions of particles may have this subtile agent for their common cause*'.[156] Faraday demonstrated that electricity was not requisite for the recombination of gases on a platinum surface, and then, after briefly relating the alternative theories put forward,[157] set about showing that the phenomena could be explained in terms of the ordinary powers of matter. This accounts for his reservation.

There are many cases of intermolecular attractions, manifested by such phenomena as surface tension, 'seeding' in crystal formation, and

---

[151] ibid., paras. 951 ff.
[152] Faraday, op. cit. (38), para. 503.
[153] ibid., entry of 1 Aug. 1832.
[154] *J. chem. Phys.* 38, 321 (1823).
[155] *Phil. Trans R. Soc.* 124, 55–76 (1834). Williams (op. cit. (44), pp. 274 ff.) has given a good account of the events leading up to this paper.
[156] Faraday, op. cit. (85), para. 564.
[157] Dulong and Thenard (*Ann. Chim. Phys.* 24, 380–7, a cautious paper) suggest that the phenomena cannot be connected with any known theory; the effects are surely not purely electric in origin. Fusinieri (*G. Fis.* 8, 259 (1825)) refers to a power 'which is neither attraction nor affinity'.

deliquescence. These actions seemed to him to be in part elective, partaking in their characters

both of the attraction of aggregation and chemical affinity; nor is this inconsistent with, but agreeable to, the idea entertained, that it is the power of particles acting, not upon others with which they can immediately and intimately combine, but upon such as are either more distantly situated with respect to them, or which, from previous condition, physical constitution, or feeble relation, are unable to enter into decided union with them.[158]

By the exertion of such an attraction, gaseous particles might cohere to a metallic surface; but the action between gas and metal would necessarily be reciprocal, thus modifying the molecular forces associated with the former. The repulsive forces between gas 'molecules' would thereby be diminished.

Remember that when a gas rests against a solid, as platina, at least half its elastic power gone on the side of the solid. Not merely a condensation in bulk to one half, but the particles necessary to elasticity by their mutual action, gone.

Refer phenomena to attraction of Particles, yet not sufficient to cause chemical combination. This probably made a general principle.[159]

With the gas 'molecules' brought close together by the attraction of the metallic surface, and their mutual repulsion diminished, combination seemed to follow naturally. Faraday was quick to point to other cases where mechanical and chemical powers seemed to overlap, since: 'Think this a subject of great consequence, for I am convinced that the superficial actions of matter and the action of particles not directly or strongly in combination are becoming daily more and more important in Chemical as well as in Mechanical Philosophy.'[160] He supported this conviction with varied illustrations, adducing his own liquefaction of chlorine under mechanical constraint; the refusal of various sodium salts to effloresce when their crystal surfaces were perfect, yielding as soon as the surface was broken;[161] and Sir James Hall's discovery that, under high pressure, calcium carbonate remained undecomposed at temperatures sufficing for its decomposition at normal pressures.[162]

The sixth series of researches, together with phenomena like these, indicated a gradual shading over from chemical to mechanical force. These generally differed in degree, but not in kind; both were manifestations of the wide powers of matter. Faraday's work on catalysis thus served to strengthen his ever-growing conviction that 'all attractions of particles may have . . . [a] common cause', explicable within a

---

[158] Faraday, op. cit. (85), para. 624.   [159] Faraday, op. cit. (38), paras. 1069, 1066.
[160] ibid., para. 1109.   [161] Faraday, op. cit. (85), para. 656 (note).
[162] Various papers: *Nicholson's J.* **4,** 8, 56 (1801); **9,** 98 (1809); **13,** 328, 381 (1806); *Trans. R. Soc. Edinb.* **5,** 43 (1805); **6** 71 (1812).

dynamical system of matter. The metastability of various explosive compounds provided further support for the correlation of mechanical and chemical powers, whose 'balance' was 'beautifully held' in the case of fulminating silver. '—instead of look at F.S. as a torpid mass rather represents half a dozen sprites chained together and struggling for liberty— . . . a link between chemical and mechanical forces.'[163]

Granted the identity of chemical and electrical forces, and the polarity of the latter, a like polarity of chemical forces was indicated. The regular arrangements evinced by crystal forms showed that there was also a crystalline polarity, and it seemed clear to Faraday's unifying mind that there must be a connection between crystalline and chemical polarity. He had looked for it for some years;[164] in 1833, while working on the relation between electrical conductivity and physical state, he wondered, 'Will not white hot diamond conduct? If so, may perhaps crystallize carbon at white heat by power of the voltaic battery.'[165] The suggested experiment, and others like it, failed. Then, in 1835, Whewell wrote from Cambridge, expressing the same conviction[166]— there absolutely must be a connection between the two polarities. Faraday quite agreed

about the importance of the relation of crystalline and chemical polarity— but do not pretend to know any theory about it at present though I suspect that it will all burst forth in its true simplicity and beauty some day shortly to some of those who now think the subject worth considering.[167]

He had long sought light on the problem, from his earliest lectures at the City Philosophical Society in 1816.[168] In 1824 he attempted to ascertain the polarity of crystals;[169] his fourth series of experimental researches in 1833 was largely although obliquely aligned towards the demonstration of this polarity. Again he met with no direct success.

From 1834 to 1838, in his ninth to fourteenth series of experimental researches in electricity,[170] Faraday worried away at the problems of electrical induction and conduction, expressing his results clearly and in detail, but interpreting them in a manner completely baffling to his contemporaries. Conduction was always preceded by induction in curved lines, and contiguous particles could be a measurable distance apart. In the course of these researches, Faraday experienced the value of a dynamical theory of matter in directing his enquiries, basing his

[163] R. Instn MS. 23: lecture of 31 May 1845.

[164] Presumably since Davy's paper following the combustion of the diamond. See Chapter 2, and note (168) below.   [165] Faraday, op. cit. (38), para. 732.

[166] Letter of 11 Dec. 1835: Trinity College, Cambridge MS.

[167] 9 Jan. 1836: Trinity College, Cambridge MS.

[168] Instn elect. Engrs. MS. p. 48.   [169] Faraday, op. cit. (38), vol. 1, p. 167.

[170] Phil. Trans. R. Soc. (1835–8); Faraday, op. cit. (85), paras. 1048–1748.

papers on the implicit assumption of such a theory. Contemporary scientists, being unaware of this preconception, were generally as mystified as they were admiring. Tyndall, for example, wrote:

And then again occur, I confess, dark sayings, difficult to be understood . . .
    The meaning of Faraday in these memoirs on induction and conduction is, as I have said, by no means always clear: and the difficulty will be most felt by those who are best trained in ordinary theoretic conceptions.[171]

What, after all, were sober scientists expected to make of a viewpoint that assented to the question, 'Do not all bodies act where they are not?'?[172] Yet Faraday gave no explanation; when tackled directly, he claimed, 'I have formed no decided notion.'[173]

Assuming that Faraday adhered, even heuristically, to some brand of force atomism, one is faced with the problem of his reticence. A partial explanation has been sought in the philosophical allegiances of British scientists.[174] I suspect that a large part of the answer lies much nearer at hand. Faraday was determined to keep his researches 'strictly experimental'. His experiments suggested that matter was in some sense dynamical—hence his eccentric understanding of 'contiguous particles' —but all this was vague enough for him to deny any 'decided notion'. Whatever his ideas about matter in the early 1830s—I believe that they were certainly dynamical—they were necessarily conjectural, thus needing no public acknowledgement. That, I suggest, is why none was forthcoming before 1837. The real question is, why did Faraday *publicly support point atomism* in 1837 and 1844?[175]

Besides his private, temperamental inhibitions, there were public obstacles to the acceptance of point atoms, arising from their lack of properties. Atoms, if they are to serve any explanatory purpose, must have properties different from and fewer than those of the massive aggregates with which experience deals. Point atoms satisfy these requirements, but, being devoid of properties, are literally inconceivable, except as mathematical constructions. English common sense was unwilling to countenance the notion of these 'unreal' entities.[176] We have already encountered Daubeny's hostility to point atomism; Donovan, while he was prepared to consider the theory, admitted that the student might find it impossible to 'accept opinions revolting to his ordinary habits of thought';[177] and Davies Gilbert explosively opined that 'The

---

[171] Bence-Jones, op. cit. (3), vol. 2, p. 85: quoted from J. Tyndall, *Faraday as a Discoverer*. London (1868).   [172] See note (30) above.
    [173] *Phil. Mag.* **17**, 64 (1840).   [174] e.g. Williams, op. cit. (44), p. 78.
    [175] Williams (op. cit. (44)—see his index, *atomism*, for page refs.) gives an extended and enthusiastic answer.
    [176] Exemplified by the popularity of T. Reid's philosophy.
    [177] M. Donovan, *Chemistry*, p. 40, 4th edn, London (1839).

most ingenious baseless fabrick that ever was reared is, in my opinion, that constructed by Boscovich.'[178] It is scarcely surprising that Faraday hesitated in the face of so united an opposition.

Faraday, during the 1830s, gradually gained confidence in the theory of point atomism as he understood it. This part of the story has been enthusiastically and admirably described elsewhere.[179] In 1837 he made his first public avowal of confidence in the theory.[180] In the same year he reacted excitedly to Mossotti's paper 'On the Forces which regulate the Internal Constitution of Bodies'.[181] Mossotti assumed that between any two bodies there were three forces—he visualized material bodies as being surrounded by their electric fluid (rather like Dalton's circum-ambient caloric). There resulted an attraction between each body and the fluid of the other, the mutual repulsion of the two electric fluids, and the mutual attraction of the two bodies. Faraday was excited because, although Mossotti used fluids, his work, if true, implied the mutual identification of the attractive forces of electricity, aggre-gation, and gravitation.[182] Faraday's notes for his Friday evening dis-course on Mossotti's paper end, however, with the reflection:

How much knowledge or facts left untouched by the theory—as change of form by heat—crystallization and cohesion—all chemical affinity even evolu-tion insulation and conduction of electricity itself.

Now want experimental proof of the general law—*Mosotti's words* not impossible—change of volume by heat and cold a connexion—crystallization another—

Progress of Knowledge—not in floods—dangerous as floods of water—but a calm and dignified progress—

Nature of a thing the answer both of the ignorant and the philosopher—search for laws.[183]

He has come full circle, and is again concerned with the relation of crystalline and chemical powers.

Induction consisted in the adoption of a forced polar state by the particles. It seemed probably that particles of different bodies might present specific differences in this respect, 'the powers not being equally diffused though equal in quantity; other circumstances also, as form

---

[178] Magdalen College, Oxford MS. 400/46: letter to Daubeny, 26 Nov. 1831. Davies Gilbert (formerly Giddy) was Davy's early patron and mentor in Cornwall, and his successor as President of the Royal Society of London.

[179] Williams, op. cit. (44), chap. 7, pp. 283–315.

[180] *The Athenaeum*, 747 (1837).   [181] *Scient. Mem.* **1**, 448 (1837).

[182] This identification was regarded by the editor of the *Philosophical Magazine* as 'one of the most remarkable discoveries of the present era in science'. (*Phil. Mag.* (3) **10**, 320 ff. (1837). That the Newtonian editor of the *Philosophical Magazine* (Brewster) was able unreservedly to welcome Mossotti's paper illustrates the compatibility of force atomism in general with Newtonianism.

[183] R. Instn MS. 24.

and quality, giving to each a peculiar polar relation'.[184] The inference to elective affinities is immediate.

Crystals were made up of symmetrically placed particles. Faraday suspected that the study of crystals might reveal any anisotropy or polarity of molecular forces. Once again, however, his experiments failed to show any relation between electrical and crystalline polarity.[185] He continued to search for the relation; he investigated magnetic polarities, but, as late as 1845, the longed-for result remained hidden, in spite of his continuing persuasion that 'crystalline structure was connected with electrical forces'.[186] At last, in 1848, he wrote to Whewell, 'You remember our talk about the connexion which ought to exist between crystalline and electric forces. Well it is beginning to appear.'[187] The results were announced in the twenty-second series, 'On the crystalline polarity of bismuth (and other bodies), and on its relation to the magnetic form of force'. Once again, there is already an admirable account of the researches leading up to this triumph,[188] so that I shall merely indicate the results as they relate to the general topic of affinity.

Faraday investigated crystals of arsenic, antimony, and bismuth, and found that they all went from stronger to weaker parts of a magnetic field. He found also that crystals, suspended horizontally, aligned themselves 'parallel to the resultant of magnetic force passing through the crystal'. This was the 'magnecrystallic' force. The attribution of alignment, resulting from neither attraction nor repulsion, to a force, was unconventional, but in accord with Faraday's meaning.

I dare say I have myself greatly to blame for the vague use of expressive words . . . What I mean by the word [force] is the *source* or *sources* of all possible actions of the particles or materials of the universe: these being often called the *powers* of nature when spoken of in relation to the different manners in which their effects are shewn.[189]

He could not conceive of his results

in any other way than by a mutual reaction of the magnetic force, and the force of the particles of the crystal on each other: and this leads the mind to another conclusion, namely, that as far as they can act on each other they partake of a like nature; and brings, I think, fresh help for the solution of that great problem in the philosophy of molecular forces, which assumes that they all have one common origin.[190]

[184] Faraday, op. cit. (85), para. 1687.   [185] ibid., paras. 1689 ff.
[186] R. Soc. MS. Herschel vii, 184; Faraday to Herschel, 13 Nov. 1845.
[187] Trinity College, Cambridge, MS., 7 Nov. 1848.
[188] Williams, op. cit. (44), chap. 9, 10.
[189] Letter of 1857 to Clerk Maxwell: Cambridge University Library Add. MS. 7655/II, 14. cf. the cautious and broad definition proposed by Boscovich (*Theoria philosophiae naturalis* (Venice, 1763 edn), trans. J. M. Child, paras. 8 and 9. Chicago (1922).)   [190] Faraday, op. cit. (53), para. 2562.

Plücker had observed the repulsion of the optical axis of some crystals by a magnetic field,[191] and this also seemed to Faraday to be related to crystalline structure. Indeed, all researches on magnetism, as on other molecular forces, served only to bring them all into closer relation with one another. Magnetism

is found . . . to possess the most intimate relations with electricity, heat, chemical action, light, crystallisation, and, through it, with the forces concerned in cohesion; and we may, in the present state of things, well feel urged to continue in our labours, encouraged by the hope of bringing it into a bond of union with gravity itself.[192]

Thus, by 1848, Faraday had brought all physical powers except gravity into union with one another, as different polar manifestations of a single original. Chemical affinity was thereby established as simply a manifestation of the powers of matter, capable of shading over into magnetic, electrical, mechanical, or crystalline powers. As his ideas on affinity progressed, his ideas on the nature of matter and of force became ever more purely dynamical, and this inevitably reacted on his concept of affinity.

A summary of the development of Faraday's ideas about the nature of matter will facilitate the correlation and interpretation of his remaining researches.

Faraday and matter

Faraday's first scientific lectures were those of a Newtonian, who was aware of dynamical extensions of Newton's ideas—both by Boscovich, and by a peculiar interpretation of the 'Queries' to the *Opticks*. Until 1837, however, there is nothing in the Faraday corpus to indicate anything more specific than Newtonian dynamism, a term covering a very wide spectrum of hypotheses. In 1837, Mossotti's paper appeared in *Scientific Memoirs*; Babbage wrote to Faraday, canvassing a theory of point atoms;[193] Hamilton reported that Faraday regarded the notion of matter currently accepted by scientists as more a hindrance than a help to the prosecution of natural philosophy;[194] and Faraday publicly concurred with Hamilton's presentation of point atomism to the British Association.

His contemporaries, however, remained simply puzzled. Tyndall was not alone in failing to discern a thread uniting Faraday's researches and

[191] Williams, op. cit. (44), pp. 413 ff. Correspondence of Faraday with Plücker, Instn elect. Engrs MSS.
[192] Faraday, op. cit. (85), para. 2614.
[193] R. Instn MS. 24.
[194] R. P. Graves, *Life of Sir W. R. Hamilton*, vol. 2, pp. 95-6. 3 vols., London and Dublin (1882-9).

complaining that the researches on induction and conduction showed 'looseness' and sometimes even inaccuracy in their phraseology.[195] Yet there was a current of force atomism giving coherence to all his work.[196] This was accompanied by an insistence on polarity, following logically from the conviction that force could be neither created nor destroyed. There was a fixed amount of 'force' in the world, and the 'creation' of, say, a positive charge could only occur if accompanied by the simultaneous and related creation of a corresponding negative charge.[197] Polarity was revealed and extracted from its concealment in unity, and this necessarily applied to all forces: since matter and force were largely identified within a dynamical scheme, molecules polarized as wholes.[198]

In addition to the emphasis on polarity, there was one further aspect of Faraday's force atomism that would have surprised Boscovich. Boscovich's atoms existed by virtue only of their relations with other particles; it was this relation that called the 'powers' into being. Faraday's atoms, however, existed individually and independently, as did their aggregates; whether a particle was a single point atom or a molecular aggregate of point atoms was, for him, irrelevant to its existence as a centre of active powers. Thus Faraday wrote in a lecture notebook of 1840, 'The acting particles are not altered in themselves only associated [in chemical combination] but the permanent effect of their association is not adhesion but a change in property.'[199] Just how Faraday was able to persuade himself in 1844 that his atoms, presumably compounded from Newtonian and Boscovichean gleanings, were Boscovichean, is impossible to say; it is, however, worth noting that, since he could not read Latin, he could have had no direct access to the *Theoria*.

Robert Hare of the University of Pennsylvania had been puzzled by Faraday's idea of contiguous particles being separated from one another while remaining contiguous,[200] and Faraday's reply was not calculated to ease his puzzlement.[201] It was not until 1844 that Faraday published his *Speculation touching Electric Conduction and the Nature of Matter*, and came out firmly for point atoms as he understood them. The next month produced a short paper[202] in which Faraday seems to have set out, for his own consideration, the arguments for point atomism. This paper suggests several considerations; those concerning the theology of nature and the significance to be attached to Faraday's seemingly ambiguous use of the word 'force' have already been indicated. Also

[195] Tyndall, *Faraday as a Discoverer*, pp. 80–90. 5th edn, London (1893). See also R. Phillips to Babbage, 4 Sept. 1832 (Brit. Mus. Add. MSS. 37187, f. 113).

[196] Williams (op. cit. (44), chap. 7) presents the argument in terms of point atoms.

[197] Faraday, op. cit. (85), paras. 1627 ff.   [198] Faraday, op. cit. (38), paras. 4571–3.

[199] R. Instn MS. 23: lecture of 1840.   [200] *Phil. Mag.* **17**, 44–54 (1840).

[201] ibid., **17**, 54–65.   [202] See note (71) above.

striking is his rapid and considerable advance in dynamism beyond the position of his public speculation. In addition, there is the distinction between laws or powers innate to matter, and laws or powers impressed on matter.[203] In the manuscript paper, Faraday says that he cannot imagine 'matter without force', yet speaks of 'forces impressed upon matter': on what did God impress forces? In a lecture of the same year, Faraday asked whether radiation might not be 'power without matter'.[204] The whole maze would appear to indicate some confusion and vagueness in his ideas, suggesting that, even while advocating point atomism, he was not entirely confident of the details of the hypothesis, and was worrying away at the problem of the nature of matter.

One critic has argued that Faraday's atomism could not be Boscovichean, because of the anomalous position in which it left gravity.[205] Faraday was, from the beginning, convinced of the unity of all the forces and powers in nature.[206] In this chapter it has been suggested repeatedly that strong support for this conviction came from his natural theology.

Boscovich's force law, involving interaction between point atoms independent of surrounding particles, was compatible only with Faraday's view of gravity, and not with his view of the nature of other forces.[207] Curiously, some nineteenth-century commentators who could not dispense with the notion of imponderable fluids reached the same conclusion from quite different premises.[208]

Faraday, perhaps intuitively feeling rather than intellectually realizing this anomaly, gradually shifted the emphasis of his theory to lines of force.[209] One could still know matter only by its forces, but now these were conceived as residing in lines of force, connecting particles and also masses of matter together. Whereas before the forces had constituted the matter, now 'particles of matter' are presented as being distinct from 'lines of forces'; the former are certainly more than the points of intersection of the latter. While effecting this conceptual transfer of force to lines traversing all space, Faraday made even more of the importance of the unity of all forces[210]—all were merely different

[203] Although in practical terms this distinction is spurious, it appears not to have been so to the theologically sensitive—see, for example, A. Koyré, *From the Closed World to the Infinite Universe*, p. 274. Baltimore (1968).   [204] R. Instn MS. 23.

[205] J. Brookes Spencer, *Archs Hist. exact Sci.* **4**, 184 (1967).

[206] See, for example, Faraday, op. cit. (85), para. 2146.   [207] See note (205) above.

[208] R. Hare, *Phil. Mag.* (3) **26**, 602 (1845): '. . . . according to the speculations of Faraday, all the powers of matter are material . . . while of all these material powers only the latter [gravitation] can be ponderable.'

[209] The first important paper indicating this shift was 'Thoughts on ray-vibrations', *Phil. Mag.* **28**, 345–50 (1846). Directional 'forces' (e.g. magnecrystallic) would have reinforced this.   [210] See, for example, Faraday, op. cit. (38), para. 7872.

manifestations of one. The particles, however, were still distinct, so that affinity remained elective.

The transfer of the forces of matter from the particles themselves to lines of force joining them was accompanied by a growing realization of the importance of space. Faraday had formerly regarded space as somehow a part of matter, since the latter's attractive forces, partially constituting the matter, extended to an infinite distance.[211] In 1850, however, he warned against confusing the particles acting on each other with the space across which they acted. '. . . space comports itself independently of matter, and after a different manner.'[212] The polarity of matter was transferred to the polarity of lines of force,[213] and it was these lines, rather than individual atoms, that now represented a determinate and unchanging amount of force.[214] Boscovichean atomism, so-called, lies far behind. Interesting in this context is a letter written to Faraday by Sir Benjamin Collins Brodie in January 1859, where the latter alludes to Faraday's agreement, 'long ago', with the opinion that the hypotheses of both Daltonian and Boscovichean atoms were nothing more than 'a contrivance for bringing these things down to the level of our limited comprehension'.[215] Faraday had made such 'contrivances' useful in extending his experimental researches.

Affinity and the conservation and correlation of force

In 1835 Faraday pursued his identification of chemical and electrical power; in an ideal battery, all the chemical power would circulate and become electricity.[216] Then, as he began his work on induction, he immediately sought to apply it to electrolytic phenomena. Electrolytic solutions conducted electricity. Induction, a molecular or particulate phenomenon, preceded all conduction, therein holding the key to an understanding of electrolysis. It set the compound particles of electrolytes into a polar state of tension, producing decomposition; Faraday viewed it as

that which lifts the current forward a step: then the resolution of the particle of water compensates for that lift and set[s] all in a state, upon which a freshly exposed particle of zinc and water acts . . .
[Polarization of compound particles by induction] is in fact a virtual separation of the bonds which unite its elements, by the use of antagonistic bonds . . .
. . . each particle transfers an equal proportion of electricity, and . . . the

---

[211] It was precisely this to which Lieut. E. B. Hunt objected; how could all the matter and all the force in the universe be interconnected? *Am. J. Sci. Arts* **18**, (2) 237–49 (1854).
[212] Faraday, op. cit. (85), 2789.   [213] ibid., paras. 3072, 3074.
[214] ibid., para. 3073.   [215] Instn elect. Engrs MS.
[216] Faraday, op. cit. (53), para. 1120.

electro-chemical equivalents represent the weights of the atoms of bodies. This may be a distant, but it is a direct, relation established between the attraction of Gravity and chemical attraction.[217]

Induction was to unify all molecular phenomena; somehow gravitation had to be included in this embrace, and Faraday eagerly seized on even the remotest connections. He went on to establish the duality of electrical force—it was impossible to charge matter with just one kind of electricity, and Erman's concept of unipolarity (which Berzelius had adopted) must therefore be a fiction.[218]

That there is no absolute charge of matter, either conducting or non-conducting, is very important in relation to chemical affinity.

That which appears to be definite about the particles of matter is probably their power of assuming a *particular* state and to a *certain* amount. Compare corpuscular forces in their amount, i.e., the forces of Electricity, Gravity, chemical affinity, cohesion, etc., and give if I can expressions of their equivalents in some shape or other.[219]

In his twelfth series of experimental researches he illustrated the electrolytic mechanism proposed above; in the case of water, the molecule was polarized by induction, leading to the separation of oxygen and hydrogen. These carried in opposite directions the force they acquired during polarization. Subsequent reassociation of hydrogen with a new 'atom' of oxygen, both substances being polarized, constituted discharge.

Faraday considered that the different initial intensities required by different electrolytes for decomposition were directly proportional to the chemical affinities of the substances concerned. This very general relationship between physical state and electrical conductivity 'draws both their physical and chemical relations so near together, as to make us hope we shall shortly arrive at the full comprehension of the influence they mutually possess over each other'.[220]

He pursued this line of thought in his fourteenth series of experimental researches, and extended the analogy between chemical and electrical polarity.[221]

It was Berzelius, I believe, who first spoke of the aptness of certain particles to assume opposite states when in presence of each other. Hypothetically we may suppose these states to increase in intensity by increased approximation, or by heat, etc., until at a certain point combination occurs, accompanied by such an arrangement of the forces of the two particles between themselves

[217] Faraday, op. cit. (38), paras. 3455–9.
[218] Faraday, op. cit. (85), paras. 1635 ff. See Erman, *Ann. Chim. Phys.* **61**, 115 (1807).
[219] Faraday, op. cit. (38), paras. 3773, 4213, 4216.
[220] Faraday, op. cit. (85), paras. 1354, 1358.
[221] *Phil. Trans. R. Soc.* **128**, 265–82 (1838).

as is equivalent to a discharge, producing at the same time a particle which is throughout a conductor.[222]

The ability to assume this excited state was, according to Faraday, a primary characteristic of matter. *Elective* affinity arose because particles of a given type assumed a special state, and acquired force—polar tension—to only a determinate degree.[223] From this picture, with its emphasis upon the interrelation and arrangement of particles and forces, he went on to develop a chemical theory, based upon chemical affinity, of current electricity. There were two stages: in the first, particles approach one another, inducing a mutual polarization of the type $Zn \overset{\delta - \delta +}{\ldots} \overset{\delta -}{O} \overset{\delta +}{-} H$. In the second stage, O leaves H and combines with Zn. Current is produced in the first stage, while the second, 'by terminating for the time the influence of the particles which have been active, allows of others coming into play, and so the effect of current is continued'.[224]

Faraday's concern with chemical affinity continued unabated. In 1840 he gave a course of seven lectures at the Royal Institution on 'the force usually called chemical affinity'.[225] His conviction that chemical and electrical forces were identical, and the power of the pile seemingly limitless, made it appear that the latter could provide 'such explications of the nature of chemical affinity as will . . . ultimately . . . enable us to reverse altogether the chemical forces and so teach us how to apply them in a way and for purposes at present unthought of'. In this connection, he continued his work on the source of power in the voltaic pile—a problem whose immediacy had grown, not decreased, since he first approached it.[226] His commitment to the chemical theory was now absolute, for he had found an argument which seemed irrefutable—if contact electromotive force had any existence, 'it must be a power not merely unlike every other natural power as to the phenomena it could produce, but also in the far higher points of limitation, definite force, and finite production'.[227]

This was yet another example of the tacit assumption of the principle of the conservation of energy in advance of its explicit statement. The assumption of the existence of contact force implied not a conversion but an actual creation of power, denied the equality of cause and effect, and allowed perpetual motion. 'If the contact theory is true

---

[222] Faraday, op. cit. (85), para. 1739.
[223] ibid., 1739.
[224] ibid., paras. 1741–3.
[225] R. Instn MS. 23.
[226] Faraday gives a brief history of the debate between proponents of the chemical and contact theories in op. cit. (85), paras. 1797–1806. (Full references are given.)
[227] ibid., para. 1798.

then may the perpetual motion also be true on principle, and it would not be difficult even to make one acting mechanically.'[228] Faraday could adduce more than the usual casual arguments in favour of the conservation of energy; only God could create, and God was rational—otherwise natural theology would be without foundation.

We find no remainders or surplusage of action in physical forces. The smallest provision is as essential as the greatest, none can be spared.[229]

To admit, indeed, that force may be destructible, or can altogether disappear, would be to admit that matter could be uncreated; for we know matter only by its forces.[230]

In 1857 Faraday gave a lecture at the Royal Institution 'On the Conservation of Force'.[231] In this, he once again expressed his conviction 'that the great and governing law is one'. He also revived his argument against the absolute need of a mathematical training for the pursuit of science. Mathematics were impotent in discovery. For example, the mathematical treatment of experimentally discovered static electricity failed to reveal dynamical or current electricity.

Under these circumstances, a principle, which may be accepted as equally strict with mathematical knowledge, comprehensible without it, applicable by all in their philosophical logic whatever form that may take, and above all, suggestive, encouraging, and instructive to the mind of the experimentalist, should be the more earnestly employed and the more frequently resorted to when we are labouring either to discover new regions of science, or to map out and develop those which are known into one harmonious whole; and if in such strivings, we, whilst applying the principle of conservation, see but imperfectly, still we should endeavour to see, for even an obscure and distorted vision is better than none. Let us, if we can, discover a new thing in *any shape*: the true appearance and character will be easily developed afterwards.[232]

Behind this lecture, giving it its consistency, lay Faraday's natural theology; the *Emporio Italiano* perceived that there was more to his science than empiricism, but identified this higher source with higher Idealism.

[H]e . . . recognises both the existence and the necessity of a science superior to experimental and mathematical sciences which knows *a priori*, and seeks after the absolute principles of things; in other words, recognises what all profound thinkers have recognised, that is, an ideal and absolute science . . .[233]

[228] Faraday, op. cit. (38) para. 5231; cf. op. cit. (53), para. 2073.
[229] Faraday, op. cit. (85), para. 2968.
[230] *Proc. R. Instn Gt. Br.* **2**, 352 (1854–58).
[231] ibid., 352–65.   [232] ibid., 364–5.
[233] *Emporio Italiano*, no. 3, 1 May 1857.

They were both right and wrong, of course. Faraday had indeed a higher reference frame and guide for his laboratory work; but it was the framework of intellectual belief, and not of Platonic, Kantian, or Hegelian Idealism.

In the same lecture on conservation, Faraday appealed to chemical affinity as an instructive and suggestive illustration of the principle.

The indestructibility of individual matter, is one case, and a most important one, of the conservation of chemical force. A molecule has been endowed with powers which give rise in it to various chemical qualities, and these never change, either in their nature or amount. A particle of oxygen is ever a particle of oxygen,—nothing can in the least wear it. If it enters into combination and disappears as oxygen,—if it pass through a thousand combinations, . . . it is still oxygen with its first qualities, neither more nor less. It has all its original force, neither more nor less . . .

Again, the body of facts included in the theory of definite proportions, witnesses to the truth of the conservation of force; and though we know little of the cause of change of properties of the acting and produced bodies, or how the forces of the former are hid amongst those of the latter, we do not for an instant doubt the conservation, but are moved to look for the manner in which the forces are, for the time, disposed, or if they have taken up another form of force, to search what that form may be.

Chemical affinity is at last, like all other forces, swallowed up by and dissolved into the wider unifying concept of interconvertible forces, whose conversion is strictly regulated by the principle of conservation. Thus chemical affinity loses its individual identity in the great flux of cosmic forces.

4

ELECTRICAL THEORIES OF AFFINITY: THE INFLUENCE OF DAVY AND FARADAY: DYNAMICAL IDEAS OF AFFINITY

DAVY AND FARADAY were among the main contributors to theories of chemical affinity in the first half of the nineteenth century. Many chemists and chemical philosophers were influenced by their partly dynamical ideas, or by the wholly dynamical ones of Kant or of later German Idealists. All these sources led to related ideas on chemical affinity, which will be considered in this chapter.

The influence of Davy and Faraday

Lavoisier thought that caloric had an affinity for ponderable bodies. When the role of electricity in chemical reactions was found to be in some respects analogous to that attributed to caloric, it was easy to extend the analogy and say that the electrical fluid had an elective faculty, preferentially attracting some bodies. Many chemists did say so,[1] among them Murray, who stated as a fact needing no further elucidation that 'Light and the Electrical fluid, are chemical agents, analogous in several respects to caloric.'[2]

The general opinion at the turn of the century seems to have been that, although electricity was a chemical agent, affinity was a peculiar power. Berthollet's work on chemical affinity served in England only to point out how very complicated the study of affinity was;[3] few were persuaded that chemical affinity was not the fundamental chemical force. Thus Thomas Thomson, having read Berthollet's work, complained:

It were to be wished that chemical phenomena could be referred to a few general laws, and shewn to be the necessary result of these laws. It were to be wished that we knew the nature of these laws so precisely, as to be able to foretell before hand the changes which result from the mutual action of bodies in every particular circumstance . . .
[This has not yet been completely achieved]: partly owing to the difficulty

[1] e.g. Aldini, *Ann. Chim. Phys.* **39**, 107 (an viie); ibid., **21**, 122.
[2] J. Murray, *Elements of Chemistry*, vol. 1, p. 170. 2 vols., Edinburgh (1801).
[3] C. L. Berthollet, *Recherches sur les lois de l'affinité*. Paris (1801). For other references, see Partington, *A History of Chemistry*, vol. 4, 576. London (1961–4).

of the subject, and partly to the unaccountable negligence of the greater number of chemists who have been more anxious to ascertain particular facts than to investigate general principles, and who have often seemed to look upon general principles as altogether foreign to their science.[4]

Nevertheless, an understanding of affinity was vital to an understanding of chemistry. Thomson accordingly devoted more than a hundred pages of his work to the problem,[5] yet he scarcely mentioned electricity in relation to chemistry—at this date he presumably thought it irrelevant to the study of chemical affinity.[6]

Thomson's attitude, however, was unfashionable. The excitement generated by the discovery of the galvanic pile ensured the continued interest of chemical philosophers in electrolytic decompositions. The fashion for galvanic studies was such that a *Société Galvanique* was founded in Paris in 1802.[7] Nicholson's *Journal* became the main channel in Britain for a constant stream of papers about the pile and its effects. Volta wrote to J. C. Delametherie, arguing that electrical effects were the primary causes of oxidation in the pile, and that the liquids between the plates 'do not properly augment the electric force; —not at all;—they merely facilitate the passage, and leave a freer course for the electric fluid, being much better conductors than simple water . . .' Nicholson, however, pointed to Davy's pile, constructed with just one metal and three different fluids, and suggested that Volta's theory had been too hastily adopted; chemical effects might well be the chief agency in the pile.[8] Governor Pownall could not understand what all the trouble was about—for him, there was a fluid, the Newtonian ether, which one could take as 'a datum, as a known fact', and this was the cause of 'the phaenomena of heat, light, electricity, and all the attractions of natural or chemical affinity, and of the various interchanges therein'.[9]

Davy's first Bakerian Lecture went some way towards the demonstration of this underlying unity, and Thomas Young's *Lectures*, published in the following year (1807), lent very cautious support to Davy's conjecture about the identity of chemical and electrical forces.[10] A similar conclusion was implicit in Grotthuss's paper of 1807, 'De l'influence de l'électricité galvanique sur les végétations métalliques',[11]

[4] T. Thomson, *A System of Chemistry*, vol. 3, pp. 132–3. 4 vols., Edinburgh (1802).

[5] ibid., vol. 3, pp. 132–247.

[6] Thomson later published a separate textbook dealing with the topics of heat and electricity—but here too electrical ideas about affinity were played down. (*An Outline of the Sciences of Heat and Electricity*. London (1830).)

[7] M. Crosland, *The Society of Arcueil*, p. 183. London (1967).

[8] *Nicholson's J.* N.S. 1, 135 (1802).   [9] *Phil. Mag.* **18**, 155.

[10] T. Young, *A Course of Lectures on Natural Philosophy and the Mechanical Arts*, vol. 1, p. 684. 2 vols., London (1807).   [11] *Ann. Chim. Phys.* **63**, 5 (1807).

and two years later Avogadro provided explicit support for the idea. Work continued throughout Europe on the phenomena of galvanism. In Cuvier's review of 1810, galvanism was presented as the branch of physical science that had occasioned the most researches and the greatest efforts in the first decade of the nineteenth century.[12]

There were, of course, dissident voices raised against the identification of chemical affinity with electrical attraction. Davy had talked about electricity, and about electrical energies; but those who believed that electricity was a fluid could not accept it as simply the cause of chemical affinity.[13] Others were prepared to think in terms of electrical powers, yet still unwilling to admit Davy's hypothesis. Maycock, for example, opposed it by experimental evidence, supported by fallacious reasoning,[14] and concluded, 'In the present state of our knowledge, it would, perhaps, be most prudent to abstain from all speculations, considering the cause of attraction and repulsion, and to consider them both as properties of matter, prevailing under different circumstances.'[15]

In spite of all objections, it would appear that Davy's hypothesis (which he repeatedly qualified, but which his contemporaries seemed incapable of understanding in anything but its most extreme form) was silently gaining ground. In 1815, Thomson, while expressing his own reservations, gave it as 'the prevailing opinion of chemists that chemical affinity is identical with electrical attraction'.[16]

Alternative theories to Davy's were, however, proposed,[17] and as the latter received no additional support, its initial glamour eventually began to wear thin. Davy was proud to claim that twenty years of progress in chemistry had not necessitated any change in his theory;[18] he should have added that the same period had lent neither stronger confirmation nor greater precision to it. In the fifth edition of his *System*, Thomson, reliable indicator of competent opinion, said that Davy's electrical theory of affinity, although ingenious and plausible, was not sufficiently supported by facts to carry conviction.[19] J. Bostock echoed this when he surveyed the field of galvanic science in 1818, noting that 'much discordance of opinion still exists upon the subject, and that some strong objections attach to every hypothesis which has yet been proposed'.[20] The problem remained open, engendering debate

[12] G. Cuvier, *Rapport historique sur les progrès des sciences naturelles depuis 1789*, p. 44. Paris (1810).

[13] J. A. de Luc, *Nicholson's J.* **26**, 113 (1810).   [14] *Nicholson's J.*, **29**, 12 (1811).

[15] ibid., p. 19.   [16] *Ann. Phil.* **5**, 8 (1815).

[17] e.g. M. Donovan, *Essay on Galvanism*. Dublin (1816).

[18] *Phil. Trans. R. Soc.* **116**, 383–422 (1826).

[19] T. Thomson, *A System of Chemistry*, vol. 3, p. 20. 5th edn, 4 vols., London (1817).

[20] J. Bostock, *An Account of the History and Present State of Galvanism*, p. 153. London (1818).

and frustration. 'Can the truth or falsehood of the opinion that chemical and galvanic forces are the same, be proved?'[21] The answers throughout the literature were varied, although individuals generally knew their own minds, and fell into two main camps—one favouring fluid theories of electricity, the other favouring force theories.[22]

The latter was in rough accord with the notion that electrical and chemical forces were intimately related, and, disbelieving in the existence of an imponderable fluid, was slowly gaining adherents. In the mid 1820s electrical theories of affinity settled into a rather dull period of consolidation, with the single notable exception of Ampère's elaboration of his electrodynamical theory. It would be easy, but not very useful, to go on recounting the multitude of superficially diverse but fundamentally similar hypotheses evolved to account for affinity, within the broad framework established by Davy and Faraday. The progress of the acceptance of their ideas on affinity may be skeletally indicated by brief reference to various editions of popular textbooks.

The 1827 edition of Edward Turner's popular *Elements of Chemistry* presented affinity as 'the basis on which the science of chemistry is founded . . . the first and leading object of [the chemist's] study'.[23] Turner regarded affinity as a specific power, distinct from the other forces that acted upon matter, and for this reason, and also because of his belief in electrical fluids, he considered that electrochemical theories of affinity were at best premature.[24] Furthermore, his complete acceptance of Dalton's atomic theory made Davy's views on the powers of matter seem unnecessarily complex.[25] Turner had experimental objections also to the electrochemical view of affinity; but the theoretical objections outlined above epitomize the contrary arguments that could be raised in the 1820s.

Thomas Thomson continued to feel uneasy about Davy's theory; in 1830, however, in spite of his adherence to fluid theories of electricity, he did accept the rather curious electrochemical theory propounded by Pouillet.[26] Thomson was clearly undecided, and in the seventh edition of his *System* described chemical affinity as an unknown force.[27] Nevertheless, it appeared to him that the electrical theory 'has been embraced

[21] Prize essay set by the Royal Academy of Science and Belles Lettres of Brussels in 1819; see *Q. J. Sci. Arts* **6**, 148 (1819).

[22] Instn elect. Engrs MS.: Letter from A. de la Rive to Faraday, 15 April 1855. 'Au fond la grande différence entre vous et moi, c'est que je suis tout *moléculaire* et que vous êtes tout *force*.—Je crois à un principe passif aussi bien qu'à un principe *actif*.'

[23] E. Turner, *Elements of Chemistry*, p. 99. Edinburgh (1827).

[24] ibid., p. 84.

[25] ibid., p. 129.

[26] T. Thomson, *An Outline of the Sciences of Heat and Electricity*, p. 532. London (1830). For Pouillet's views, see *Ann. Chim. Phys.* **35**, 401 (1827).

[27] T. Thomson, *A System of Chemistry*, vol. 1, p. 31. 7th edn, 2 vols., London (1831).

PLATES

Plate 1

Humphry Davy

Plate 2

Michael Faraday

Plate 3

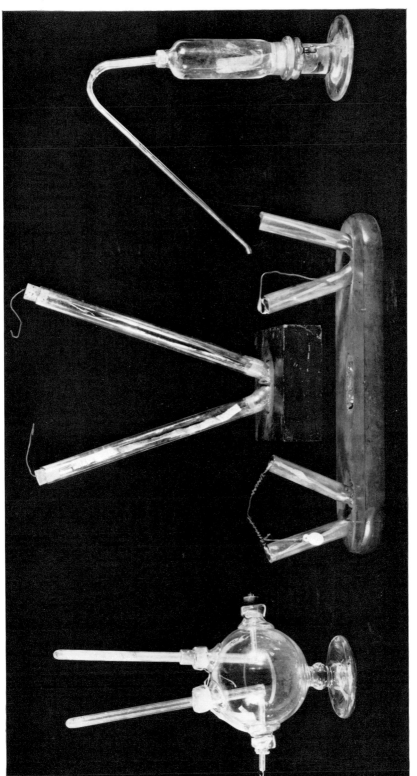

Faraday's electrochemical apparatus

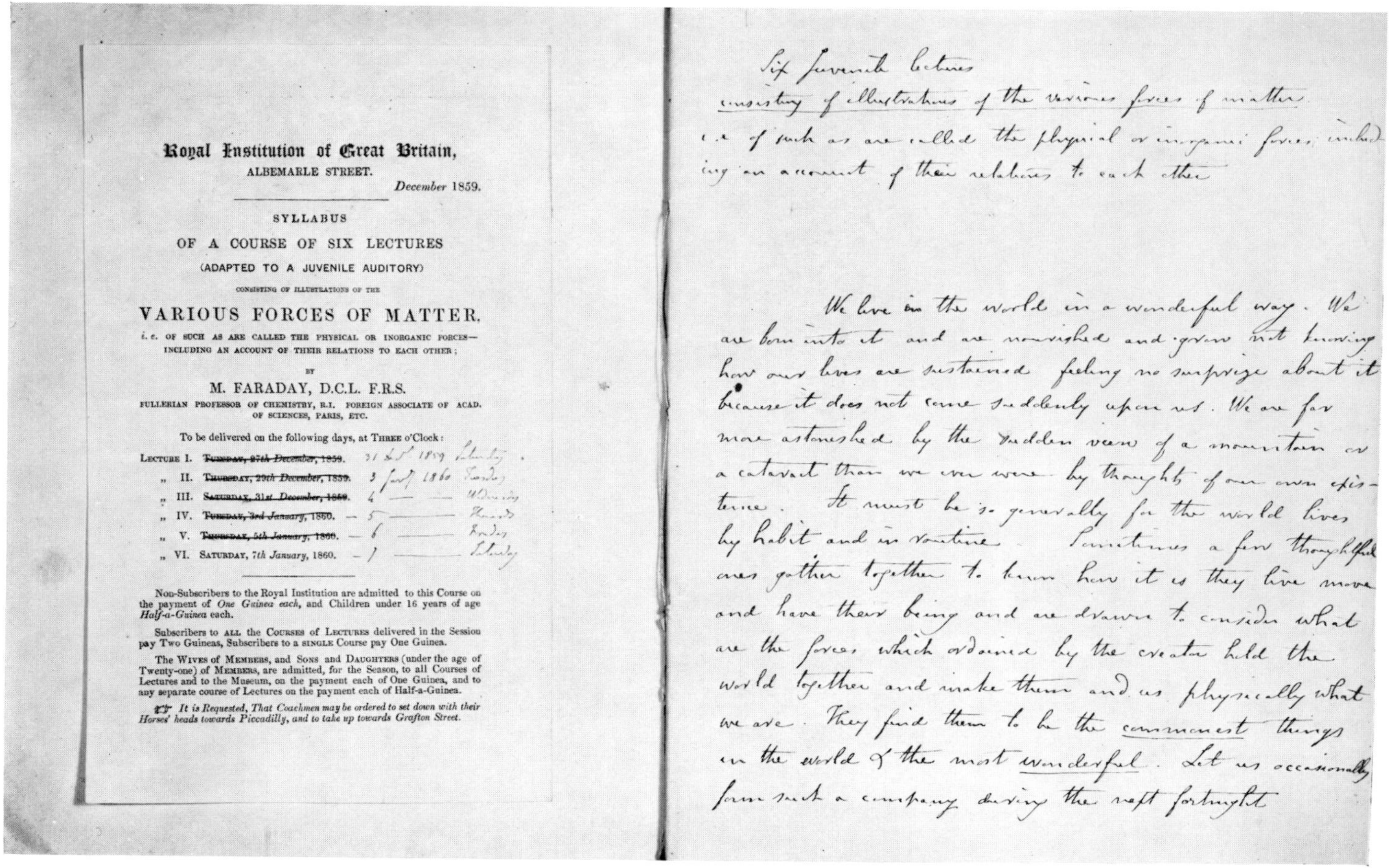

Royal Institution of Great Britain,
ALBEMARLE STREET.
December 1859.

SYLLABUS
OF A COURSE OF SIX LECTURES
(ADAPTED TO A JUVENILE AUDITORY)
CONSISTING OF ILLUSTRATIONS OF THE
VARIOUS FORCES OF MATTER,
i.e. OF SUCH AS ARE CALLED THE PHYSICAL OR INORGANIC FORCES—INCLUDING AN ACCOUNT OF THEIR RELATIONS TO EACH OTHER;
BY
M. FARADAY, D.C.L. F.R.S.
FULLERIAN PROFESSOR OF CHEMISTRY, R.I. FOREIGN ASSOCIATE OF ACAD. OF SCIENCES, PARIS, ETC.

To be delivered on the following days, at THREE o'Clock:

LECTURE I. ~~TUESDAY, 27th December, 1859.~~ 31 Decr 1859 Saturday
„ II. ~~THURSDAY, 29th December, 1859.~~ 3 Jan 1860 Tuesday
„ III. ~~SATURDAY, 31st December, 1859.~~ 4 — Wednesday
„ IV. ~~TUESDAY, 3rd January, 1860.~~ — 5 — Thursday
„ V. ~~THURSDAY, 5th January, 1860.~~ — 6 — Friday
„ VI. SATURDAY, 7th January, 1860. — 7 — Saturday

Non-Subscribers to the Royal Institution are admitted to this Course on the payment of *One Guinea each*, and Children under 16 years of age *Half-a-Guinea* each.

Subscribers to ALL the COURSES of LECTURES delivered in the Session pay Two Guineas, Subscribers to a SINGLE Course pay One Guinea.

The WIVES of MEMBERS, and SONS and DAUGHTERS (under the age of Twenty-one) of MEMBERS, are admitted, for the Season, to all Courses of Lectures and to the Museum, on the payment each of One Guinea, and to any separate course of Lectures on the payment each of Half-a-Guinea.

☞ *It is Requested, That Coachmen may be ordered to set down with their Horses' heads towards Piccadilly, and to take up towards Grafton Street.*

Six Juvenile Lectures
consisting of illustrations of the various forces of matter
i.e. of such as are called the physical or inorganic forces; including an account of their relations to each other

We live in the world in a wonderful way. We are born into it and are nourished and grow not knowing how our lives are sustained feeling no surprize about it because it does not come suddenly upon us. We are far more astonished by the sudden view of a mountain or a cataract than we ever were by thoughts of our own existence. It must be so generally for the world lives by habit and in routine. Sometimes a few thoughtful ones gather together to know how it is they live move and have their being and are drawn to consider what are the forces which ordained by the creator hold the world together and make them and us physically what we are. They find them to be the commonest things in the world & the most wonderful. Let us occasionally form such a company during the next fortnight

Faraday's lectures on the various forces of matter

Plate 5

André-Marie Ampère

Plate 6

William Whewell

Plate 7

Hans Christian Oersted

Plate 8

BARON BERZELIUS, M.D., F.R.S.

LATE PROFESSOR OF CHEMISTRY IN THE UNIVERSITY OF STOCKHOLM &c&c

Jöns Jacob Berzelius

Plate 9

J.–B. Dumas

Plate 10

Auguste Laurent

Plate 11

5 cm

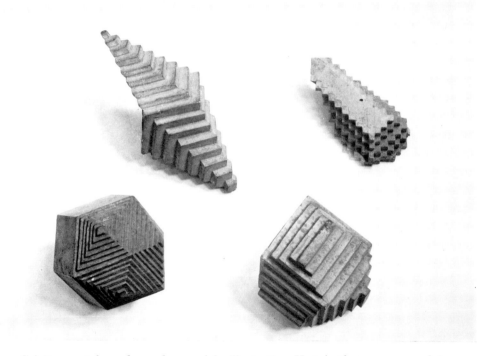

Calcite crystals and wooden models illustrating Haüy's theory: prepared by Pleuvin and Journy (photographs by G. L'E. Turner)

Plate 12

Charles Gerhardt

Plate 13

Claude Louis Berthollet

Plate 14

M. Berthelot

by every chemist of eminence, and seems at present to constitute a universally admitted opinion'.[28]

By 1836, Mary Somerville was able to announce that 'Dr. Faraday has *proved*, by recent experiments, . . . that chemical affinity is merely a result of the electrical state of the particles of matter.'[29] There had been no mention of affinity in the index to the first edition of her *Connexion of the Physical Sciences*.[30] The strong statement of proof made in the third edition was a consequence both of Faraday's experimental researches (fifth, seventh, and eighth series), and of her correspondence with Faraday in 1833–4.[31] If Mrs. Somerville's work, in its successive editions, is used as a barometer of the general climate of educated scientific opinion in England,[32] it suggests that Faraday's work served to revive interest in electrochemical theories, which had made no real advance in England since Davy's early and dramatic discoveries. The fourth edition of Brande's *Manual of Chemistry* (1836) casually tacks the implications of Faraday's work on to the modified Newtonianism of the earlier editions;[33] and the very unconcern with which this is done implies complete acceptance of what had ceased to be controversial.

Faraday's rejection of the possibility of contact electricity removed one of the mainstays of Davy's electrochemical theory, which had as its chief recommendation the reconciliation and synthesis of the chemical and contact theories of the origin of voltaic electricity. As Dumas wrote in 1837, Davy's theory of affinity satisfied all the data of chemistry, but came up against the radical difficulty of its incorporation of the contact theory. 'So we cannot admit the truth of Davy's theory, and yet it is great and beautiful. It was adequate for all the phenomena of chemistry, it sufficed to provide its author with a brilliant career.'[34]

In succeeding years, work on the correlation of forces in general inevitably had implications for chemical affinity, uniting it more firmly with the general matrix of the forces of nature. Joule's early work established that the heat of combustion was an electrical effect,[35] and in 1843 Liebig wrote that there was a quantitative correlation between chemical

[28] ibid., vol. 1, p. 38.

[29] M. Somerville, *On the Connexion of the Physical Sciences*, p. 123. 3rd edn, London (1836).

[30] ibid. London (1834).

[31] Bodleian Library, Somerville papers, vi: Faraday to Mrs. Somerville, Nov. 1833 and Aug. 1834.

[32] E. Patterson seems to agree that this is a reasonable use. 'Mary Somerville', *Br. J. Hist. Sci.*, **4**, 311–39 (1969).

[33] W. T. Brande, *A Manual of Chemistry*, p. 50. 1st edn, London (1819), 4th edn (1836).

[34] J.-B. Dumas, *Leçons sur la philosophie chimique recueillies par M. Bineau*, p. 409. Paris (1837).

[35] *Rep. Br. Ass. Advmt Sci.* p. 31 (1842); *Phil. Mag.* (3) **22**, 204 (1843).

affinity and electricity.[36] In the second edition of his *Introduction to Chemical Philosophy*, Daniell extended his earlier views and identified chemical affinity with both current affinity and electricity.[37]

Baudrimont went so far as to say that this and similar identities removed the need for the concept of chemical affinity as a distinct force. In his *Traité* he does say that chemical affinity is a force in its own right,[38] but the subsequent discussion makes nonsense of this statement, and suggests that he put it in merely to define it away.[39] Baudrimont recognized, as few of his contemporaries did, that a general theory of chemistry would have to impinge upon general physics to a considerable extent, and would have to take into account the essential properties of matter.[40] Different aspects of his work will therefore be considered in later chapters. His geometrical and crystallographic concepts will be of importance in discussing organic chemistry and affinity, while his dynamical speculations were prophetic of later developments in chemistry, including chemical thermodynamics.

In the 1850 edition of his *Elements of Chemistry*, Thomas Graham stated that chemical polarity was the fundamental concept;[41] chemical and electrical forces were so intimately related that he was unwilling to suggest which might be original.[42]

In 1853 Gerhardt opened the way to a reconciliation of severely modified versions of the dualist and unitary theories in a paper on acid anhydrides.[43] He argued that substances were not absolutely electronegative or positive, and should all be ranged in a continuous series indicating their electrical relations.[44]

After this date, electrical theories of affinity were of little interest in themselves; chemistry was aiming at a wider perspective, wherein structural and mechanical factors became increasingly important.[45] Electrical theories of affinity attracted little attention in the 1850s, and indeed these years saw a temporary lull in debates on the nature of chemical affinity. Structural concepts were being developed, together with thermochemistry. Yet even the latter, while it offered a promising approach to the measurement of affinity, said very little about its nature.

Wurtz, in 1869, made the timely admission that all theories of

[36] J. von Liebig, *Familiar Letters on Chemistry*, p. 38. London (1843).

[37] J. F. Daniell, *Introduction to Chemical Philosophy*, pp. 462 ff. 2nd edn, London (1843).

[38] A. E. Baudrimont, *Traité de chimie générale et expérimentale*, vol. 1, p. 200. 2 vols. (1844–6).   [39] ibid., vol 1, p. 216.   [40] ibid., vol. 1, p. 283.

[41] T. Graham, *Elements of Chemistry*, vol. 1, p. 238. 2nd edn, 2 vols., London (1850).

[42] ibid., vol. 1, p. 244.   [43] See Chapter 6 below.

[44] *Ann. Chim. Phys.* **37**, 285 ff. (1853). Gerhardt was, however, more interested in formal homologous series than in electrochemical series.

[45] See Chapters 6 and 7 below.

affinity were then premature; the full extent of the problem was becoming apparent. Affinity was clearly connected with electricity—but the theory of electricity itself was lacking.[46] The extent to which ignorance was recognized is suggested by a passage from the sixth edition of Grove's *Correlation of Physical Forces*, which contains a whole chapter on chemical affinity. In spite of nearly a century of research, and in spite of the undoubted connection between affinity and electricity: 'CHEMICAL AFFINITY ... is that mode of force of which the human mind has hitherto formed the least definite idea.'[47]

Ampère

André-Marie Ampère's electrodynamical theory owes little to the work of Davy and Faraday, in spite of many points of contact. It is, however, both original and important in the history of affinity, and deserves particular attention.

Born in 1775, Ampère was a close contemporary of Humphry Davy, corresponding with him on a variety of chemical topics. In 1812 Ampère informed Davy of Dulong's perilous discovery of nitrogen trichloride,[48] while two years later Ampère was responsible for providing Davy with a sample of newly discovered iodine,[49] to the extreme irritation of other French chemists. Whereas Ampère had formerly been most concerned with the philosophy of science,[50] his long conversations with Davy turned his thoughts firmly towards chemistry.[51]

Two related problems had arisen in the course of their discussions together: the possibility of quantification in chemistry, and the nature of matter.[52] These problems fascinated Ampère, and he quickly began work on a paper discussing the application of mathematical considerations to the physics of gases and of crystallization, using both chemical evidence and Haüy's crystallography.[53] Ampère wanted to complete his paper before Davy left Paris; but the problem grew in scope as he advanced, and his lectures at the *Institut* were a constant drain upon his time and energy. He made rapid progress, however, and by March 1814 was within sight of a break-through.

[46] A. Wurtz, *Dictionnaire de chimie pure et appliquée*, vol. 1, p. 78. 3 vols. (1869–78).

[47] W. R. Grove, *The Correlation of Physical Forces*, p. 137. 6th edn, London (1874).

[48] *Correspondance du grand Ampère*, ed L. de Launay, vol. 2, pp. 416–17. (3 vols., Paris, 1936–43): Ampère to Davy, 26 Aug. 1812. See H. Davy, *Collected Works*, ed. J. Davy, vol. 5, pp. 391–7. 9 vols., London (1839).

[49] H. Davy, op. cit. (48), vol. 5, pp. 437–56.

[50] See his MSS. in the archives of the *Académie des Sciences*, and also his *Essai sur la philosophie des sciences*. Paris (1834).

[51] Ampère to Roux, 11 March 1814: quoted in de Launay, op. cit. (48), vol. 2, p. 463.

[52] op. cit. (48), vol. 2, pp. 454–63.

[53] For a detailed account, see S. H. Mauskopf, 'The Atomic Structural Theories of Ampère and Gaudin', *Isis* **60**, 61–74 (1969).

For two months now I have been working on a project whose results seemed to me to open a new career in [chemistry], and to provide the means for *a priori* prediction of the fixed proportions according to which bodies combine, by referring their different combinations to principles which would be the expression of a law of nature, whose discovery will perhaps be the most important idea I have had, after my achievements in metaphysics last summer. I say 'after what I have done in metaphysics', because the latter science is the only really important one: for the theory of chemical combinations is otherwise clear and incontestable, and will become as common-place as other theories generally accepted in physical science.[54]

Davy encouraged Ampère in his efforts,[55] which bore fruit in a paper in the *Annales de chimie*.[56] Starting from an independent re-statement of Avogadro's hypothesis that equal volumes of gases contained the same number of molecules, and making only the simplest assumptions, he concluded that 'elementary' particles were in fact compound, and of determinate primitive form.[57] Molecules of oxygen, nitrogen, and hydrogen, for example, were each composed of four 'particles' arranged at the corners of a regular tetrahedron, while chlorine molecules consisted of eight 'particles' at the corners of an octahedron. Then, using the structural relations between these figures, he built up composite molecules still corresponding in shape to Haüy's primitive forms. Only certain combinations of these units were geometrically possible, so that the relative combining proportions of bodies were restricted and determined.

It follows from what we have just said, that when particles combine to form a single particle, it is by arranging themselves so that the centres of gravity of the constituent particles coincide; the corners of one particle are located in the intervals left by the corners of the other, and conversely. It is in this way that I picture chemical combination, and this is where it differs from the aggregation of similar particles, which takes place by simple juxtaposition.

The assumption that structural and crystallographic factors were adequate for the explanation, and even for the quantitative prediction, of chemical combination, was adopted and elaborated in his lectures and manuscript notes.[58] A form of point atomism emerges here; Ampère developed his theory by treating the fundamental particles as points united by their central forces. He was, however, unwilling to reject imponderable fluids, and in his lectures of 1815 spoke of caloric, electric, and magnetic fluids.[59]

[54] See note (51).
[55] J. A. Paris, *The Life of Sir Humphrey Davy, Bart.*, vol. 2, p. 34. 2 vols., London (1831).
[56] *Ann. Chim. Phys.* **90**, 43–86 (1814).   [57] See note (53) above.
[58] *Académie des Sciences*, Paris MS. Ampère xiii, 239, 240.   [59] ibid. MS. Ampère 238.

There seems to have been little development of these ideas until 1820, when Oersted's discovery of electromagnetism was announced. The initial reaction in Paris was cold[60]—but then it was found that the phenomenon was constant and reproducible, whereupon 'everyone flung themselves upon it, and now, at every meeting of the *Institut*, new experiments on the subject are all they talk about'.[61] Biot was out of town just then, and Ampère made the most of his opportunity. He had been strongly influenced by Coulomb, who asserted that electricity and magnetism were distinct fluids:[62] yet Ampère enthusiastically welcomed Oersted's 'beautiful discovery'.[63] Ampère's acquaintance with Davy and with Kant's philosophy may perhaps have partly prepared him for electromagnetism.

Electricity acted upon magnetism. This might suggest the fundamental identity of the two phenomena, in which case electromagnetism could be interpreted as the action of electricity upon itself.[64] The next step would be to investigate the possible interaction of two current-carrying wires. Whether or not Ampère's reasoning followed this course, he did carry out the latter investigation, found the interaction he had hoped for, and went on to develop the mathematical and experimental implications of this discovery.[65] He proposed the theory that magnetism was caused by the motion of positive and negative electric fluids around and perpendicular to the axis of the magnet. His interpretation had obvious weaknesses.[66] What, for example, was the origin of the currents in permanent magnets?[67] Berzelius, who like everyone else at the time was busy conducting experiments on electromagnetic and electrodynamic phenomena, rejected Ampère's conclusions,[68] and

[60] See de Launay, op. cit. (48), vol. 2, p. 566. Ampère to Roux-Bordier sen., 21 Feb. 1821. See also L. P. Williams, *Michael Faraday*, pp. 142–3. Chapman and Hall, London (1965).

[61] Söderbaum, ed. *Jac. Berzelius Bref*, vol. 4, p. 17 (6 vols. + 3 suppl., Uppsala, 1912–32): Dulong to Berzelius, 2 Oct. 1820.

[62] See Williams, *op cit.* (60), pp. 142–3.

[63] de Launay, op. cit. (48), vol. 2, p. 562.

[64] See note (62) above.

[65] Ampère, *Mémoires sur l'électrodynamique*, vol. 1, pp. 48 ff. and vol. 2, p. 1. 2 vols., Paris (1885–7).

[66] Williams points this out, giving a fuller statement of Ampère's conclusions (op. cit. (60), p. 144).

[67] e.g. Berzelius to A. Marcet, 20 Nov. 1821: Söderbaum, op. cit. (61), vol. 1, p. 224.

... comme nous ne connaissons point d'autre courant électrique que celui qui se produit pour refaire l'équilibre des forces électriques, comment est-ce que ce courant se ferait sans cesse dans la substance homogène du fer ou de l'acier? Est-ce que l'on peut considérer comme une explication admissible une hypothèse qui a besoin d'adopter des propriétés de l'électricité que nous ne lui connaissons pas?

[68] See note (67) above and Berzelius to Berthollet, 19 Dec. 1820: quoted in Söderbaum, op. cit. (61), vol. 1, p. 74.

Faraday too had serious doubts.[69] In Geneva, 'tout est Ampère', but in London, Wollaston, with his physical rather than mathematical bias, was sure that such facile acceptance of an improbable doctrine was wrong. There was a great want of experimental evidence.[70]

Fresnel soon persuaded Ampère to give up his coaxial currents, and to replace them by molecular currents.[71] These implied an entirely new theory of matter, and correspondingly strange ideas about chemical processes in general and chemical affinity in particular. The theory gained wide acceptance, in spite of its novelty and hypothetical nature.

The fullest statement of Ampère's theory of molecular currents, and of its application to the problem of chemical combination, seems to be a 67-page manuscript in the archives of the *Académie des Sciences*.[72] Ampère began with the luminiferous ether required by Fresnel's undulatory theory of light.[73] Ampère considered the ether, a neutral fluid spread throughout all space, as being composed of the two electricities. Their separation from the ether gave rise to all electrical phenomena. Mutual attraction between electrically opposed particles, and repulsion between like particles, led to a series of compositions and decompositions that were propagated through the surrounding fluid in a direction perpendicular to the original decomposing influence.[74, 75] The atoms or particles of ponderable matter had an unchanging inherent electricity, which circulated around their centres,[76] attracting the opposite fluid from the surrounding neutral ether, and repelling the like component until equilibrium was restored. Such actions could account for chemical combination when the interatomic distances involved were sufficiently small for the opposite electricities from the disturbed ether to recombine.[77] The principal stages in combination are illustrated in Fig. 4.1.

Different elements would have different electrical atmospheres, and during chemical combination all the atmospheres of the less electrified

---

[69] Williams, op. cit. (60), pp. 147 ff. See also Faraday to Marcet, 15 Jan. 1822 (R. Instn MS. 11, fl. 1): '. . . though M. Ampère seems very well satisfied with his theory in its present state yet certainly much more must be done with it and much more proved by experiment before the cautious philosopher will receive it into his stock of knowledge or think it at all approaching to invulnerable.' Apart from distrusting the lack of experimental evidence, Faraday would have been reluctant to accept a theory based on imponderable fluids.

[70] See note (61) above.

[71] Ampère, op. cit. (65), pp. 145–7; *Contemp. Phys.* **4**, 113 (1962).

[72] *Académie des Sciences*, Paris MS. Ampère xiii, 250/8.

[73] ibid., p. 1.

[74] ibid., pp. 1–2; cf. (65), vol. 1, p. 216.

[75] For a technical treatment of Fresnel's work, see E. Whittaker, *A History of the Theories of Aether and Electricity*, vol. 1, pp. 116 ff. 2 vols., Harper, New York (1960).

[76] Ampère, op. cit. (65), p. 2.

[77] ibid., pp. 7, 11–14.

particles would tend to reunite with an equal quantity of the opposite
electricity from the atmospheres of the more highly electrified ones,
thus forming neutral ether. This reunion constituted chemical com-

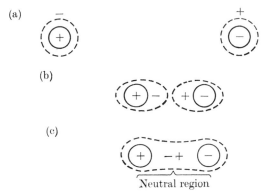

(a) (b) (c)

Fig. 4.1. (a) The atoms are still too far apart to affect one another. (b) The
atoms are now close enough for mutual polarization to occur. (c) Combination
has taken place.

bination. In the special conditions presented by the voltaic pile, the
separated electricities passed through the metallic conductor prior to
recombination, manifesting themselves as a voltaic current.[78] (See
Fig. 4.2.) Chemical combination might thus produce a current. In

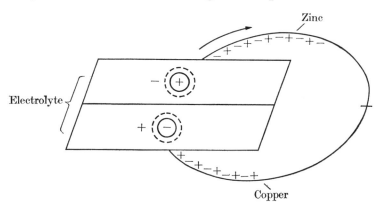

Fig. 4.2.

general, however, the recombination of positive and negative electricities was direct. When, for example, hydrogen and oxygen combined, no electricity appeared, because the small excess of the electrical
atmospheres of each gas remained in the water vapour, forming neutral
ether.[79]

[78] ibid., pp. 12, 58.   [79] ibid., p. 14.

Ampère's theory had the particular merit of explaining combination between bodies of similar electrical natures, since it made polarization of electrical atmospheres dependent on the relative electrical states of atoms. Thus sulphur and oxygen were both permanently electronegative, but to different degrees.[80] They could therefore combine. Another virtue of the theory was that it enabled Ampère to reveal and maintain his intuitive grasp of the principle that chemical combination occurs only when energetically favoured. This principle seems, at first sight, to rule out combinations between particles with the same inherent electricities, for example, between chlorine and oxygen.[81] But such reactions did occur.

Consider the combination of two bodies, A and B, of electricities $+a$, $+b$. The compound AB will have electricity $+(a + b)$, while the surrounding atmosphere will have electricity $-(a + b)$. However, because of the repulsion between like electricities, the atmosphere will be unsymmetrical and polarized, concentrating around the particles of lesser inherent electricity. This favours combination between particles of like electricity, 'because the mutual repulsion of the particles of the more intense atmosphere is more easily satisfied by the extension of this atmosphere'.[82]

Ampère concluded his exposition by noting, almost as an afterthought, that the heat and light evolved in chemical reaction could also be explained in terms of decomposition and recombination in the neutral electrical fluid or luminiferous ether.[83]

His theory was certainly comprehensive; but it involved a number of *ad hoc* assumptions about the nature of matter in general and of electricity in particular, and was therefore unacceptable to his more cautious contemporaries and also to less cautious ones whose *ad hoc* assumptions differed from his. Berzelius believed that, even after Oersted's discovery, scientists were still a long way from knowing the connection between electricity and magnetism,[84] and was not at all satisfied with Ampère's views. 'If a hypothesis needs to adopt properties of electricity which are unknown to us, can we consider it as an acceptable explanation?'[85] Berzelius nevertheless hoped that Ampère's

---

[80] Although bodies had specific inherent electrical states, they normally failed to reveal them; being in contact with conductors (presumably through the action of the decomposable ether) they attracted opposite electrical atmosphere and therefore presented a neutral aggregate.

[81] This example is the author's, not Ampère's.

[82] Ampère, op. cit. (65), pp. 54–5: 'parce que la répulsion mutuelle des particules de l'atmosphère la plus intense se trouve mieux satisfaite par l'extension de cette atmosphère'.

[83] ibid., pp. 60–3.

[84] *Jahresber. Fortschr. Chem.* **1**, 1–2 (Tubingen, 1822).

[85] See note (67) above.

experiments would soon lead to discoveries about the nature of electricity and about 'the particular nature of molecules or atoms'. Faraday maintained his old reservations about Ampère's ideas,[86] although Ampère excitedly seized upon Faraday's electrodynamical researches as providing perfect confirmation of his theory.[87]

The theory did, however, have undoubted advantages. Ampère's explanation of the combination of similarly electrified bodies was superior to Berzelius's, while Davy failed to provide any explanation. Ampère's theory gave also a reasonable explanation of the fact that bodies remained in combination, without drifting apart, after their electricities had mutually neutralized one another.

His ideas met with some approval. Auguste de la Rive, for example, employed Ampère's electrodynamic molecular model in his explanation of electrolysis.[88] In England, however, the theory remained little known and less accepted. The *Quarterly Review* tried to remedy this situation by an enthusiastic account. Ampère

seems to have constructed the master-key which is adapted to open every compartment of this intricate science, and procure us a clear and consistent view of the whole. As the theory to which we allude does not appear to have attracted, in this country, the attention it deserves, and as, indeed, there exists no work in English in which it is more than slightly adverted to, we shall here attempt to give . . . a sketch of its leading features.
. . . It will be found, that no theory of electromagnetism hitherto devised can at all enter into competition with that of Ampère. It is impossible to deny that a great advance will have been made in the philosophy of nature, if it can be shown, or even rendered probable, that [magnetism and electricity result from the same agency].[89]

Part of the resistance in England to Ampère's ideas arose from incomplete comprehension. Thomas Thomson, for example, condemned his electrodynamic theory because it failed to account for combination between bodies of similar inherent electricity;[90] but a plausible explanation of such combinations was one of the principal advantages of Ampère's system.

Faraday was one of the few English scientists to give any serious consideration to Ampère's theory, even though its assumption of

[86] Faraday to de la Rive, 12 Sept. 1821: quoted by Bence-Jones *The Life and Letters of Faraday*, vol. 1, pp. 354–7. 2 vols., London (1870).

[87] de Launay, op. cit. (48), vol. 2, pp. 486 ff.: Ampère to Faraday, 10 July 1822 (Instn elect. Engrs MS.); vol. 2, p. 576: Ampère to Bredin, 3 Dec. 1821. See also Faraday, *Q. J. Sci. Arts* **12**, 74–96 (1822).

[88] *Ann. Chim. Phys.* **28**, 190 (1825). Thenard, in *Traité de chimie élémentaire, théorique et pratique*, vol. 5, p. 510 (6th edn, 6 vols., Paris, 1834–6) adopted a synthetic compromise between Ampère's and Berzelius's theories.

[89] J. Murray, *Q. Rev.* **35**, 237–69 (1827).

[90] T. Thomson, *System of Chemistry*, pp. 37 ff. 7th edn, 2 vols., London (1831).

imponderable electrical fluids was totally at variance with his notion of electrical powers. He promised Ampère to read his papers as they came out,[91] and admitted that his mathematical theory was perhaps the only one to have actually predicted any new facts.[92] In 1834 Faraday noted that Ampère's theory was compatible with his own conviction that all electrical effects could be viewed as arising from induction.[93] Perhaps, if Ampère had not relied on fluids, Faraday would have had no reservations; as it was, he wrote to Whewell in 1835:

> I ought to say that I accept Ampère's theory as the best present repre-
> sentation of facts, but that still I hold it with a little reserve. This reserve is
> more a general feeling than any thing founded on distinct objections to it.
> Remember I am no mathematician. If I were one and could go into a closer
> examination of the theory than is at present possible for me I might have no
> doubts left;—but all my mathematics consist in that rough natural portion
> of geometry which every body has more or less. Hence the reason why I
> have never put my facts into terms of Ampère's theory; and why I cling to
> the relations of Magnetic and Electric forces as the simplest I can perceive;
> these again are readily distinguished in practise and hence the most con-
> venient if not the best for an experimentalist to refer to.—I wish most
> sincerely some mathematician would think it worth his while to do that for
> the facts which I can not do for them.[94]

Whewell was certainly better placed than Faraday to understand the mathematics of Ampère's theory, which he called a 'beautiful and com-prehensive generalization'.[95] He observed that it was superior to rival hypotheses and indicative of the advanced condition of physical specu-lation. The theory had, moreover, explained more than it had been designed for, thereby showing signs of truth and stability.[96] Whewell was probably influenced in favour of the theory because Ampère shared his general assumptions about electrical fluids and dynamical polarity.[97]

Faraday's twentieth series of experimental researches, 'On New Magnetic Actions, and on the Magnetic Condition of all Matter',[98] also could be used to support Ampère's claims. Faraday now claimed that the magnetic induction of particles was the cause of the magnetic

---

[91] Faraday to Ampère, 4 May 1833: Instn elect. Engrs MS.
[92] Faraday to Mrs. Somerville, Nov. 1833 and Aug. 1834: Bodleian Library MS. Somerville papers vi. An example was the prediction of strength of fields produced by constant currents.
[93] *Phil. Mag.* **5**, 349–54 (1834).
[94] Trinity College, Cambridge MSS.: Faraday to Whewell, 19 Sept. 1835.
[95] W. Whewell, *History of the Inductive Sciences*, vol. 3, p. 76. 3 vols., London (1837).
[96] ibid., vol. 3, pp. 81 ff. See also Whewell to Faraday, 9 Sept. 1835: Trinity College, Cambridge MS.
[97] ibid., vol. 3, p. 88. Whewell was to insist upon the polar character of electric cur-rents, which derived from their progressive and directional nature.
[98] M. Faraday, *Experimental Researches in Electricity*, paras. 2223 ff. London (1839–55).

affections of gross matter, and this could be interpreted as arising from currents circulating round the particles of matter.

A theory of point atomism was implicit in Ampère's geometrical explanation of chemical combination (1814). He finally published his ideas explicitly in 1835, in a note 'On Heat and Light considered as the Results of Vibratory Motion'. He began by distinguishing between particles, molecules, and atoms. 'The term *molecules* I give to an assemblage of atoms held at a distance from each other by the attractive and repulsive forces proper to every atom . . . What I call *atoms*, are the material points from which these attractive and repulsive forces emanate.'[99] He then deduced the polyhedral forms of molecules from equilibrium configurations of small groups of point atoms, corresponding to the primitive forms of Haüy's crystallography.

Faraday, meanwhile, gradually shifted the emphasis of his thoughts about the powers of matter from central forces to lines of force radiating from and passing through material atoms. He came to think of space as a matrix for a network of lines of force, and accordingly rejected his former idea that space occupied by force constituted matter. He no longer regarded each particle as extending, by virtue of its forces, to the ends of the universe; he therefore had to distinguish the particles acting upon one another from the space across which they acted. Ampère's theory postulated points and an axis about which electrical currents circulated, and Faraday dispensed with it because by 1850 he could no longer imagine geometrical constructions in a vacuum serving as centres for this rotation.[100] His general uneasy feeling of 1835 had deepened to conviction—he rejected Ampère's theory because it felt wrong—it was inconceivable, rather than contrary to experience.

Ampère's ideas received more publicity in France than in England. Dumas, who in 1837 condemned Ampère's theory as absolutely inadmissible,[101] had at first been captivated by it. His manuscript lecture notes at the *Académie des Sciences* were cautious, and admitted only that affinity seemed analogous to electricity; nevertheless, he gave a long and detailed exposition of Ampère's theory, as far as he understood it.[102] In one lecture he even defined chemistry as the science concerned with the laws and effects of molecular forces,[103] and many undated manuscript notes show that he thought of chemical processes in terms of some sort of dynamical atomism.[104] Several other French

[99] *Phil. Mag.* (3) **7**, 342 (1835).
[100] Faraday, op. cit. (98), 25th series, paras. 2718 ff., especially 2787–8. See also *Diary*, paras. 10834–42, entry of 8 April 1850.
[101] J.-B. Dumas, op. cit. (34), pp. 414–15.
[102] *Académie des Sciences*, Paris MS. Dumas vii (n.d.), 'Statique chimique'.
[103] ibid., MS. vii, p. 4, 'Affinité'.
[104] ibid., p. 7.

chemists, including Gaudin, Laurent, and Gerhardt, knew of and used aspects of Ampère's theory.[105]

Ampère himself naturally tried to introduce his own ideas into his lectures, although these necessarily stressed more conventional topics. He discussed electrochemical decomposition, which, with related phenomena,

> make[s] it impossible to doubt that this action takes place between the particles of bodies, and that it co-operates with the force of chemical affinity, if it is not itself one of the latter's principle causes. Why should the unknown cause of heat, which seems to be so closely related to that of electricity, not produce analogous results . . .?[106]

In addition, he ascribed chemical affinity, cohesion, crystallization, etc., to the action of molecular forces in polar tension,[107] but only hinted at the full scope of his ideas. His caution may have stemmed from mere pedagogical orthodoxy; it may, however, have shielded him against a widespread dislike of speculative science, a reaction against *Naturphilosophie*.

William Whewell

The influence of William Whewell on Faraday, or at least the interplay of ideas between them, has repeatedly emerged in the preceding pages. In this section, Whewell's ideas on chemical affinity will be considered.

In 1838 the *Edinburgh Review* published a summary of Comte's *Positive Philosophy*,[108] claiming that it had been well known and 'highly appreciated' in London, before the publication of Whewell's *History*. Now there was a controversy between Idealists and Positivists, harmful to the image of science.

> It is, however, some consolation that the leading combatants are not men who have added much to positive discovery; and that those who are destined to maintain the vestal fire [on the altar of the temple of Science] are not likely to disturb the flame which has been fanned by themselves.[109]

In other words, professional scientists had no time for philosophical controversy—and a good thing too! Whewell's whole career had been a standing challenge to this viewpoint, for he had lectured and written on a wide variety of topics, from moral philosophy to mineralogy. His reading was as comprehensive as his taste for generalization, which led Sir Joseph Hooker to suppose that he had read the prefaces of a good many books. He also wrote a good many. In the introduction to his

---

[105] See Chapter 6 below.   [106] *Académie des Sciences*, Paris MS. Ampère xiii, 237.
[107] ibid., 238.   [108] D. Brewster, *Edinb. Rev.* **67**, 271 (1838).   [109] ibid., p. 307.

History he declared the necessary interdependence of facts and ideas in science,[110] and ascribed to the latter an almost scandalous importance, viewed from the standpoint of nineteenth-century British empiricism.

Whewell stressed the importance of ideas in any system of scientific thought, and was himself much concerned with the ideas of unity and polarity in the physical world. A taste for unity emerged clearly at the beginning of his discussion of the 'mechanico-chemical' sciences, where he proceeded from the phenomena of galvanism to those of chemical decomposition, concluding that

[t]he highest generalisations to which we can look, in advancing from the elementary facts of electricity and galvanism, must involve chemical notions; we must therefore, in laying out the platform of these sciences, make provision for that convergence of mechanical and chemical theory, which they are to exhibit as we ascend.[111]

Following this assertion, there came a historical treatment of the principal theories of electricity and magnetism, united by the concept of polarity; Whewell considered that all this was 'only one-half of the science to which it belongs,—one limb of the colossal form of chemistry'.[112] The next task, for him and for chemistry, was to indicate the relation of electrical polarity to chemical composition. Indeed, he described the history of chemistry as being a mere interruption of the history of electrodynamical research.[113] In these terms it was natural to regard Faraday's unifying work as the latest stage in the progress of the history of chemistry, and Whewell stated that Faraday, in making his discoveries, needed to understand chemical affinity as 'a peculiar influence or force, which, acting in opposite directions, combines and resolves bodies'; such comprehension was also essential if Faraday's theories were to be understood.[114]

In his *Philosophy of the Inductive Sciences*[115] Whewell revealed to their full extent the underlying ideas that he believed accounted for the continuity of the sciences revealed in his *History*. He was firmly convinced that the continuity of ideas in the human and in the divine mind reflected a corresponding continuity and unity in the physical world. As the *Quarterly Review* noted, Whewell adopted the doctrines of Idealism to a far greater degree 'than we are at all aware of having been before done, except perhaps in the writings of some of the later German metaphysicians',[116] even though he roundly castigated the

---

[110] Whewell, op. cit. (95), vol. 1, p. 6. [111] ibid., vol. 3, p. 6.
[112] ibid., vol. 3, p. 98.
[113] ibid., vol. 3, p. 154.
[114] ibid., vol. 3, p. 172.
[115] W. Whewell, *Philosophy of the Inductive Sciences*. 2 vols., London (1840).
[116] J. F. W. Herschel, *Q. Rev.* **68**, 177 (1841).

metaphysical excesses of Hegel and Schelling.[117] Facts, as well as ideas, were after all, essential to the prosecution of science.

The guiding idea in the mechanico-chemical sciences had, according to Whewell, been that of polarity, which he admitted 'belongs peculiarly to the region of speculation'.[118] Chemical affinity, involving contrariety and mutual neutralization, implied the idea of polarity, as did the existence of regular crystalline form. Magnetic and electrical polarities provided other obvious instances. Whewell was clearly favourable towards the principle of the 'connection of polarities', which affirmed that all polarities were fundamentally identical. The principle suggested a union of electrical with chemical science. The chemical forces that held together or separated the parts of bodies were identical with the polar forces of electricity; this great discovery was now 'entirely established in the minds of the most profound and philosophical chemists of our time'. Faraday's researches had furnished the main evidence for the principle; but it was the principle itself, in all its simplicity, that persuaded the mind that the discoveries must be true.[119]

Whewell went on to consider the connection of chemical and crystalline polarities, which followed almost as a tautology from the 'fact' that the 'forces which hold together the elements of a crystal . . . are the same forces which make it a crystal'.[120] Even though chemical polarity was most easily regarded as analogous to electrical polarity, and crystalline polarity was a structural phenomenon, there was nevertheless a fundamental identity between them, as Berzelius had asserted[121]—and as Faraday was later to demonstrate to his own and Whewell's satisfaction.

Even if chemical affinity was regarded as a force subsumed in the general system of polar forces, its nature remained obscure. But something more was known; for example, experiment had shown that affinity was definite as to the kind and quantity of the elements that it united.[122] Indeed, now that these aspects of chemical affinity had been clearly recognized, it was impossible to conceive of chemical combinations of continuously variable composition. Whewell, failing to point out the circularity of this particular example of belated synthetic *a priorism*,

[117] op. cit. (115), vol. 1, pp. 356 ff. Whewell did, however, acknowledge his indebtedness to Kant (see *On the Philosophy of Discovery*, p. 335 (London, 1860); he also explained the theological basis of his philosophy.

[118] Whewell, op. cit. (115), vol. 1, p. 33.

[119] ibid., vol. 1, p. 352.

[120] ibid., vol. 1, p. 353.

[121] J. J. Berzelius, *Essai sur la théorie des proportions chimiques, et sur l'influence chimique de l'électricité*, p. 113. 2nd edn, Paris (1835).

[122] These were immediate deductions from the laws of chemical composition. Whewell, op. cit. (115), vol. 1, p. 384.

concluded that there was an intimate relation between chemical composition and affinity. He assumed not merely that the internal powers by which the ultimate particles of bodies were held together were responsible for the composition of these bodies, but also that the same powers were what determined bodies to be bodies, and were responsible for all their properties; 'for if these forces are not the cause of these peculiarities, what can be the cause?'[123] Composition and properties thus depended upon one another so that '[t]he foundation of all our speculations respecting the intimate constitution of bodies must be this, that their composition determines their properties'.[124] So keen was Whewell on this guiding principle, that he was able to instance Davy's earlier opinion, that diamond and charcoal differed in chemical properties because they also differed in composition, without even alluding to his later volte-face.

Physical properties, then, depended entirely upon chemical composition. Whewell seems to have ignored the idea of polarity here; or perhaps he so completely accepted the implications of polarity that he saw no need to make them explicit, since they were implicit in his concept of composition. The dependence of properties upon composition suggested that the more definite the former were, the more distinct would be their dependence upon the latter. Now it appeared to Whewell that crystalline structure was the most definite of all physical properties of chemical substances, and it must therefore be 'in a most intimate dependence' upon chemical composition.[125] A different chain of argument, with several doubtful links, has returned him to his starting point—the connection between chemical and crystalline polarities.[126]

If 'composition' is given its generally accepted meaning, then the claim that physical properties depend entirely upon chemical composition seems to imply the acceptance of an atomic view of matter; according to such a view, the whole is neither more nor less than the sum of its parts, and mutual modification of properties, depending upon arrangement, does not occur. Yet Whewell found Dalton's atomic theory untenable, because he could not reconcile it with the facts of analytical chemistry.[127] For example, on the basis of the molecular weights then adopted, various metallic oxides contained fractions of atoms—an absurd contradiction in terms. Furthermore, the conventional atomic theory assumed that hard atoms were somehow united by

---

[123] ibid., vol. 1, p. 385. [124] ibid., vol. 1, p. 386.

[125] ibid., vol. 1, p. 387. Whewell's stress on the importance of crystalline form, and its relation to constitution, appears also in his *Essay on Mineralogical Classification and Nomenclature, passim.* Cambridge (1828).

[126] cf. Whewell to Faraday, 11 Nov. 1848 (Trinity College, Cambridge MS.): '. . . crystalline form is determined by chemical composition'.

[127] Whewell, op. cit. (115), vol. 1, p. 408.

mechanical forces of attraction, and, as Whewell remarked, simple mechanical attraction between hard atoms is incapable of accounting for the elective nature of affinity.[128] Newtonian physics applied to chemistry was of little use. Whewell wrote in 1835:

That by considering fluid or solid bodies as composed of distinct particles, and by suitably assuming the forces which these particles exert on each other, we may represent their mechanical condition, and trace its consequences, is undoubtedly true. But it would be to go too far, to assert, on this account alone, that the only true conception of the physical structure of bodies is that which represents them as so constituted. This would be to mistake the use of the differential calculus for the evidence of a physical truth.[129]

This, together with the preference he displayed for a molecular hypothesis representing matter as a collection of molecules 'or centres of force',[130] shows that a special meaning is indeed to be attached to 'composition' as he uses it. The hopes that he held out for isomorphism as a link between chemistry and crystallography illustrate his awareness of the influence of arrangement on properties.[131]

Chemical affinity was, for Whewell, a non-mechanical polar force, to be identified with the other polarities in the physical world. Its action determined the composition of bodies, in terms of the combining proportions of their elements, even while the arrangements of molecules determined the equilibrium pattern of the interaction of molecular forces, and thus the properties of solid bodies; since the equilibrium configurations themselves were functions of the forces of chemical affinity, the latter could be regarded as the cause of all the chemical and physical properties of bodies. This is a highly generalized, and therefore somewhat vague, scheme; but it does incorporate the work of Davy, Faraday, and Berzelius, and of inorganic and physical scientists in general. It ignores, however, the new perspectives then opening in the field of organic chemistry.[132] The triple clash between Whewell's guiding ideas of unity, polarity, and structure will provide threads of coherence in the subsequent discussion of affinity and organic chemistry.

The Nature Philosophers

Whewell's concern with the relation of Kantian philosophy to physical science could not have become more extreme without losing

---

[128] See Chapter 1, note (38).   [129] *Rep. Br. Ass. Advmt Sci.*, p. 1 (1835).
[130] Whewell, op. cit. (115), vol. 1, pp. 417 ff.   [131] *Phil. Mag.* N.S. **10**, 401 (1831).
[132] Additional lacunae were pointed out by J. T. Merz, *A History of European Thought in the Nineteenth Century* (London, 1907), vol. 1, pp. 307–10; repr. 4 vols., Dover, New York (1965). For example, no hint of the conservation of energy or of the mechanical theory of heat appears—which is only to say that Whewell was essentially a reporter, not a prophet.

general respect in England: his equal concern with the brute facts of science preserved his reputation, and he was a mainstay of the Cambridge intellectual network. None of the individuals or theories described in the earlier parts of this chapter lacked scientific respectability. They often provoked disagreement, but they had to be seriously considered.

There was, however, a stream of ideas, not without influence on the physical sciences in general and on electrochemical theories of affinity in particular, which many would have judged to be without the pale of respectability. These ideas derived from *Naturphilosophie*. Now it so happens that the conclusions of the Nature Philosophers frequently coincided with those of Davy and of Faraday, so that the former regarded their premises as confirmed by the experiments of the latter[133] —even though Faraday, and, less unequivocally, Davy were fundamentally antipathetic to *Naturphilosophie*. Another distinction that should be made at the outset is between the superficially similar schemes of Kantian and post-Kantian dynamism, whose respective adherents were often bitterly opposed.[134] Temperamental as well as philosophical prejudices are at issue, giving rise to fundamentally different approaches to science; beside this, any coincidental identity of conclusions is barely significant.

In 1901, in a course of lectures on *Naturphilosophie*, Wilhelm Ostwald told a story to illustrate the way the previous generation of Nature Philosophers had thought.

An Englishman, a Frenchman and a German were given the task of describing the camel. The Englishman, so the story tells us, wanted to take his rifle, go to Africa, shoot a camel, have it stuffed, and put it in a museum. The Frenchman wanted to go into the *Jardin d'acclimatation* in Paris and study the camel there; and, if he failed to find one there, he would be inclined to have grave doubts about its existence, and would in any case allow it only very slight importance. The German, on the contrary, needed only to go to his study, and there construct the nature of the camel from the depths of his spirit.[135]

Ostwald used this story to exemplify the disastrous extension of Schelling's idea that laws of nature could be built infallibly from thought, since *Denken* and *Sein* were equivalent, and the same laws governed the spiritual and the external worlds. The remainder of this chapter will indicate some consequences of this attitude to nature, in so far as it is germane to the history of chemical affinity.

Schelling's specific ideas about chemical affinity have already been

[133] See, for example, Schelling, *Ueber Faraday's neueste Entdeckung*. Munich (1832).
[134] See, for example, G. Eriksson, *Lychnos*, 1–37 (165).
[135] W. Ostwald, *Naturphilosophie*, p. 7. 2nd edn, Leipzig (1902).

A M—K

touched upon in Chapter 2. Their influence, and the compromises that were adopted between them and their Kantian source, remain for consideration. We have seen already that Kantian and later Idealistic dynamism are incompatible, since the latter avoids and ignores the restrictions imposed by the former. If one ignored these restrictions, then it was possible to apply Kantian metaphysics to chemistry, and indeed to all the phenomena of nature, and to claim that they could all be deduced from an extension of Kant's metaphysical principles. Kant had specifically rejected this claim; chemistry, for him, was accordingly incapable of advancing beyond the state of a historical or descriptive art, and any speculations about the fundamentals of chemistry necessarily remained mere speculations. Schelling, rejecting the *Ding an sich*, deliberately rejected these restrictions on a thoroughgoing idealism; but there were many who did so tacitly and covertly, or who may, indeed, have been unaware of their very existence. Accordingly, what they claimed as Kantian metaphysics and Kantian chemistry were nothing of the sort; but because of their ignorance, whether wilful or innocent, of their alleged master's instructions, they described themselves as his followers. The historian is therefore faced with an array of 'Kantians' who were in fact Nature Philosophers. Contemporary accounts of the movement are correspondingly confused. The precise metaphysical standing of each hypothesis described in the succeeding pages will not be analysed; the comments just made should, however, be remembered.

A system of philosophy that promised that careful introspection would reveal all the truths of nature could not but win adherents, and, in the first decades of the nineteenth century, it won many, especially in Germany.[136] Among these was the philosophical chemist, Johann Wilhelm Ritter, friend of Schelling and of Schlegel, who interwove speculative *Naturphilosophie* with laboratory experiments in electrochemistry so tightly that it is hard to see which of his conclusions are philosophical and which empirically founded. In 1798, two years before the invention of the voltaic pile, Ritter 'attempted to create an electrochemical theory in which the forces of electricity and chemical affinity were treated as identical'.[137] In 1805, he gave a more extended statement of his theory.[138] All bodies could be arranged in a row, in the order of their electrical and chemical energies; and yet, paradoxically, in this row where order and place alone ruled, the different bodies were in constant strife. The whole universe was in fact in a state of dynamical

[136] *Phil. Mag.* **2**, 277 (1798).
[137] Williams, op. cit. (60), p. 62, referring to J. W. Ritter, *Beweis dass ein bestandiger Galvanismus den Lebensprocess in dem Theirreich begleitet; Nebst neuen Versuchen und Bemerkungen über den Galvanismus* (Weimar, 1798). (Ronald's Library.)
[138] J. W. Ritter, *Über das electrische System der Körper*. Leipzig (1805).

equilibrium between the contending powers of a universal duality. Such ideas were central to the tenets of *Naturphilosophie*. The other fundamental idea was that Nature's present multiplicity was once preceded by a Unity. The original passage from One to Many was the same as the one we still see today in living organisms, an expression of the tension of vitality,[139] which, being universal, is manifested in so-called dead as well as in living matter. Thus inorganic nature has an evolutionary history, quite as much as life has, and electricity, identical with chemical affinity, has a major role to play in its evolution.[140] Thus was electrochemistry reduced to order. Ritter's philosophical compatriots applauded him. Foreign scientific critics were interested in his experimental results, fascinated by and distrustful of his philosophical webs. The English in particular generally handled his papers at arm's length.

Winterl proposed a system closely related to Ritter's, founded upon two principles, acidity and basicity, or positive and negative electricity, whose mutual tendency to combine accounted for all chemical reactions. All bodies were originally composed of like atoms, and the particular characters of each of them depended upon their degree of adherence to the principle of acidity or basicity; Winterl regarded adherence itself as a third, immaterial principle.[141] As Cuvier observed in 1810, this system fitted in nicely with the metaphysical dualism fashionable in Germany, and accordingly found credit there.[142]

The founders of these *a priori* systems could claim that any discovery that agreed with their predictions was really theirs, since they had originated the idea. More conventional scientists, zealous for their own priority, were indignant at such claims[143]—hence Davy's retrospective outburst in 1826, when he unfavourably contrasted Winterl and Ritter, those profitless speculators, with scientists like himself, who proceeded upon the factual evidence elicited by their researches.[144] This attitude grew apace, even as the metaphysical system it abhorred gained strength in Germany. As early as 1800 Schiller wrote of the 'encroaching hatred of philosophy'.[145]

[139] ibid., pp. 388, 394.   [140] ibid., p. 403.

[141] J. R. Partington, *A History of Chemistry*, vol. 3, p. 600. 4 vols., London (1961–4). Ann. Chim. Phys. **1**, 175–99 (1804).

[142] G. Cuivier, op. cit. (12), pp. 64–6.

[143] cf. Whewell, op. cit. (115), vol. 1, p. 357.

It was very natural that chemists should refuse to acknowledge, in this fanciful and vague knowledge . . . an anticipation of Davy's doctrine of the identity of electrical and chemical forces, or of Oersted's electro-magnetic agency. Yet it was perhaps no less natural that the author of such assertions should look upon every great step in the electro-chemical theory as an illustration of his own doctrines.

[144] *Phil. Trans. R. Soc.* **116**, 383 ff. (1826).

[145] G. Hennemann, *Naturphilosophie in 19 Jahrhundert*, p. 43. Freiburg, Munich (1959).

One of the most popular textbooks, and in many ways the most singular one, advocating dynamism was that of F. C. Gren.[146] The translator of his *Principles of Modern Chemistry* informs us that, at the time of writing, Gren had renounced the 'Atomistical System', and had adopted Kantian dynamism.[147] The work, however, is an extraordinary and eclectic amalgam, as its author noted with pride.[148] It is also inconsistent within itself, accepting imponderable fluids on one page, and rejecting them on another.[149] As for affinity, it both was and was not an inherent power of matter, and for want of experimental evidence its laws had been vainly sought. Tables of affinities were perhaps the most useful guide in the meantime. This empirical approach to chemistry was in accord with Kant's instructions; Gren was one of the very few who seems to have tried to follow them.[150]

In spite of the missionary zeal of Coleridge, Nitsch, and a few others, there were few converts in England to the German Philosophy of Nature. The *Quarterly Review* and *Edinburgh Review* were, in the first decade of the nineteenth century, full of remarks, more or less wistful, about the utter indifference of Englishmen to any metaphysical system and about the consuming interest aroused by the experimental sciences.[151]

The first social stirring came with the visit of Mme de Staël to London, where she dined on her first night with Sir Humphry and Lady Davy. Marcet wrote to Berzelius that 'society' rushed around her. 'However, since it isn't the style of any party in this country to intrigue with women, she has plenty of leisure in which to busy herself with metaphysics, and she goes around, making, or trying to make, proselytes to German philosophy.'[152] But although Mme de Staël was undoubtedly a huge social success, there is little to suggest that her proselytizing won support for *Naturphilosophie*. British empiricism was well entrenched.

It was not until Thomson reviewed Oersted's *Ansichten der chemischen Naturgesetze* (Berlin, 1812) that the English public was confronted with even a garbled version of the scientific ideas of Nature Philosophy in a reputable domestic journal.[153] It is indicative of the lack of

---

[146] F. C. Gren, *Principles of Modern Chemistry*. 2 vols., London (1800).

[147] ibid., p. iii.

[148] ibid., p. 15.

[149] Contrast the preface with p. 102; see also Gren to Van Marum, 12 June 1791 (Hollandsche Maatschappj der Wetenschappen Van Marum Archives): '*elecy. is the material cause of combination.*'

[150] F. C. Gren, op. cit. (146), vol. 1, pp. 46–9. 2 vols., London (1800).

[151] *Q. Rev.*, **6**, 1 ff. (1811); *Edinb. Rev.* **1**, 253 (1803); **17**, 167 (1810) provide good examples.

[152] Söderbaum, op. cit. (61), vol. 3, p 63: letter of 28 July 1813.

[153] *Ann. Phil.* **5**, 5–8 (1815).

communication between Germany and England at that date that Thomson had been forced to write his review without seeing the original work; he had been able to obtain only partial secondary accounts. Oersted was strongly influenced by Schelling's philosophy; he was also one of the major scientists of the first half of the nineteenth century, and he made important contributions to chemical and electrical science.

Hans Christian Oersted: Idealistic metaphysics and chemical affinity

Oersted was born in 1777; his father was an apothecary. He studied at the University of Copenhagen, and in 1799 submitted his doctoral thesis on 'The architectonics of the metaphysics of nature', in which he not only expounded the positive content of his science (physics and chemistry), but also united these within the embrace of Kant's philosophy.[154] His thesis was the extension of an earlier paper giving a critical account of Kant's *Metaphysische Anfangsgründe der Naturwissenschaft*, and he had thoroughly imbibed Kant's view that 'a rational doctrine of nature deserves the name of natural science only when the natural laws at its foundation are recognized *a priori*, and are not mere laws of experience'.[155]

In his student years, he showed outstanding interest and ability in practical and theoretical chemistry, and also explored philosophical systems other than Kant's. In 1799 he reviewed two of Schelling's books on *Naturphilosophie*, admiring the 'great and beautiful ideas' he found in them. But he was unhappy at Schelling's insufficient care in distinguishing empirical propositions from *a priori* ones, especially because the empirical propositions adduced were often utterly false. Oersted's student years were his most nearly Kantian ones, when he believed that laws of nature were absolutely valid only when they had *a priori* foundations, and that a corpuscular atomic theory was simply illogical.[156]

After submitting his doctoral thesis, Oersted was awarded a travelling scholarship, which he used to visit France and Germany between 1801 and 1803. He spent most of his time in Germany, where he heard Fichte lecture, and defended him in conversation. He listened also to Schlegel holding forth about the poetical treatment of physics, and the latter's connection with mythology. He met Schelling too, liked his way of tackling things, but objected to his lack of first-hand knowledge of nature. He did, however, agree with Schelling that moral and

[154] Hennemann, op. cit. (145), p. 76.

[155] R. C. Stauffer, *Isis* **48**, 33–50 (1957).

[156] K. Meyer, 'The Scientific Life and Works of H. C. Oersted' in *H. C. Oersted, Scientific Papers*, vol. 1, p. xviii. 3 vols., Copenhagen (1920).

physical nature were intimately connected. This conviction underlies Oersted's remark in 1817: 'The spirit that should emanate from the whole being of the scientist should, it seems to me, be religio-poetico-philosophical; without this, he does not know the end to which the sciences strive and can never be anything but a subordinate.'[157]

Oersted's overall philosophy seems to be his own, though not too far removed from Kant's. He borrows from Schelling, yet aims to weld disparate parts into a systematic whole. Most of the parts of his system derive from other men's writings, while the synthesis is his own. This is in no way an aspersion on his intellectual integrity—as a philosopher, he was seeking truth, not priority. Fichte, Schelling, Schlegel, Schleiermacher—he met them all. But the greatest mutual influence and most lasting friendship was that struck up with Ritter, philosopher and empiric, at the hub of electrochemistry and *Naturphilosophie*. It was Ritter who tempted Oersted, for a time, to the brink of Schelling's philosophy, and Ritter again who showed how that philosophy could pay dividends in the laboratory. Oersted was entranced, and only gradually emancipated himself. This was made easier by Ritter, who gradually lost himself in speculation, and became as careless as Schelling about his facts. But in 1803, when Oersted returned to Denmark, he had no reservations about Ritter, and, like him, was much attracted by Winterl's system.[158] It seems clear from their extensive correspondence[159] that Ritter was a major stimulus for Oersted's work on electromagnetism.[160]

Once Oersted was home, his empiricism reasserted itself. In 1806 he was appointed to a chair in physics at the University of Copenhagen, and began again to carry out experimental work. In spite of his metaphysical predilections, he recognized the value of observation in science. As he wrote in 1810:

It is also my firm conviction, and my lectures bear witness thereof, that a great fundamental unity pervades the whole of nature; but just when one has become convinced of this, it becomes doubly necessary to direct one's whole attention to the world of the manifold, wherein this truth above all finds its confirmation. If one does not do this, unity itself remains an unfruitful and empty idea which leads to no true inspiration.[161]

The blend of *a priori* speculation with empiricism was, however, capable of considerable variation. Oersted, in the early part of his career, could hardly have been accused of an unreasonable bias towards observation. His 1803 summary of Winterl's *Prolusiones ad chemiam*

---

[157] K. Meyer, op. cit. (156), vol. 1, pp. xxv–xxvi.
[158] Stauffer, op. cit. (155), 37.
[159] *Correspondance de H. C. Oersted*, ed. Harding. 2 vols., Copenhagen (1920).
[160] cf. Stauffer, op. cit. (155), p. 40. [161] Cited by Stauffer, op. cit. (155), p. 39.

saeculi decimi noni (1803) was laudatory; Winterl was, in Oersted's estimation, far more profound and comprehensive in his ideas than Lavoisier had been.[162] Shortly afterwards a review appeared in France, expressing nothing but horror and indignation at Winterl's extravagant nonsense.[163] It was only gradually that Oersted shifted from these metaphysical extremes.

A conviction of unity, a blend (in varying proportions) of metaphysics with experiment, an interest in electrochemistry—all these were still shared with Ritter in 1812, when Oersted published his *Ansichten der chemischen Naturgesetze*. The book was published in a much-altered translation in Paris the following year, under the title of *Recherches sur l'identité des forces chimiques et électriques*. Coleridge was quick to obtain a copy of the German edition, and his copious notes swarm across the margins.[164]

The purpose of the work was to show that all previously investigated chemical changes could be deduced from two universally distributed natural forces, and that chemical and electrical forces were identical.[165] The complete theory necessarily required an extensive knowledge of electricity and of its effects, and accordingly, 'Ritter can be considered as the creator of the new chemistry'.[166] Davy knew the French translation of the book, and not unnaturally took exception to this claim, which ignored his own experimental contributions—the only kind of any value in science.[167]

In previous investigations of chemical phenomena, affinity or attraction had represented the ultimate limit of knowledge; but since the nature of chemical affinity remained unknown, this was a mere verbal pretence. All existing chemical systems (presumably excepting Ritter's and Winterl's) presented the subject as consisting of a multiplicity of unexplained phenomena related by an unknown force. It was necessary to emerge from this obscurity, and to seek in chemistry the same unity and completeness as that which had already been reached in mechanical science. 'Chemistry will then become a science of forces; mathematics will reach into its inmost recesses, to determine the quantitative relations, directions and effective forms of these forces.'[168] Oersted's book was to be a first step towards this emancipation. He sought quantification in chemistry, while retaining the Nature Philosophers' emphasis on dynamism, unity, and polarity.

Combustion was to serve as the beginning of his investigation, both because it manifested polarity and activity to an exalted degree, and

[162] Whewell, op. cit. (115), vol. 1, p. 356.
[163] *Ann. Chim. Phys.* **1**, 191 (1804). [164] Br. Mus. C.43a.17 (North Library).
[165] ibid., p. 7. [166] ibid., p. 11. [167] See note (144).
[168] H. C. Oersted. *Ansichten der chemischen Naturgesetze*, p. 5. Berlin (1812).

because accurate experimental investigations already existed in this branch of chemistry. Oersted's highly individual blend of different approaches to science is strikingly apparent here. The universal forces, as manifested in chemical phenomena, were combustion and combustibility; their combination and equilibration led to acidity and alkalinity. In general, chemical combination was an equilibration of these two forces, both of which were found in dynamical equilibrium in all substances. The forces were also identical with mechanical forces; the underlying unity of forces was the reason for the overall unity of nature.[169]

Coleridge, poring over such statements, was insistent that continuity and unity should be recognized as embracing and exceeding this polarity. 'The great Law, by which the Powers at their maxima pass into their opposite poles, the dilative for instance (in Warmth) into the contractive (in Fire) contains an incomparably fuller solution than the arbitrary metaphor, attraction.'[170] Oersted, however, continued to speak of two distinct forces responsible for acidity and alkalinity.[171] Both of these forces were, in themselves, expansive, but their reciprocal attraction led to a contractive effect; since all substances contained a balance of these forces, chemical combination was effected by two pairs of reciprocal attractions.[172] Oersted pointed out that the simultaneous existence of attractive and repulsive forces in dynamical equilibrium satisfied the necessary conditions for the existence of matter, so that the two forces of combustion and combustibility accounted for all chemical phenomena, all mechanical action including heat, and the very existence of matter. The question of the original nature of these forces, however, was best left to pure speculative philosophy.[173] The implied severance of pure speculation from empirically based speculation would have made little sense to extremist Nature Philosophers.[174] For Coleridge, however, Oersted did not go far enough in divorcing the two.

[169] op. cit. (168), pp. 75 ff.

[170] ibid., marginal note p. 77. Coleridge objected also to Oersted's classification of all bodies into combustible, combustive, and neutral (alkalis, acids, salts)—note p. 1: 'I cannot with my present knowledge of chemistry see the advantage of this division and it strikes me as an objection to it that it confounds the substantiative Forces with the modifying, by directing the whole attention to the latter.' This unification was for Oersted one of the prime advantages of his system.

[171] cf. Winterl.   [172] op. cit. (168), pp. 109–10.   [173] ibid., p. 115.

[174] The mean position of *Naturphilosophie* was frequently exaggerated, both by its opponents and by its more enthusiastic supporters. Schelling's philosophy included a 'speculative physics', which claimed to construct natural forces from first principles. It did not, however, claim to reveal empirical facts for the first time, but merely to exhibit the pattern in Nature (i.e. Nature as first known empirically); it sought to reveal the why and the wherefore of the facts. cf. F. Copleston, *A History of Philosophy*, vol. 7, pp. 135 ff., 138. Image, New York (1965).

It is of [the] highest importance in all departments of knowledge to keep the Speculative distinct from the Empirical. As long as they run parallel, they are of the greatest service to each other: they never meet but to cut and cross. This is Oersted's fault—the rock of offence on which this Work strikes. Davy is necessarily right: for he follows the established *Regula recta* of empirical chemistry, *viz.* that all bodies shall be considered as simple, till they have been *shewn* to be compound. On this Rule, Chlorine, and Iodine claim the title of simple Bodies (*Stoffen*) with the same right as Oxygen, or the Metals, while the Speculative Chemist sees *a priori*, that all alike must be composite.[175]

Oersted's dynamical theory implied that there were no true elements; it is curious to find Coleridge in empirical mood criticizing his own firmly held beliefs. Indeed, Coleridge committed in his marginal notes the very solecism of which he accused Oersted.[176]

By considering the voltaic pile as a circle, and the associated phenomena as a circulation of force, Oersted was able to claim as a logical deduction that chemical and electrical forces were identical;[177] internal changes of bodies were called chemical, while the external consequences of the concurrent redistribution and re-equilibration of forces led to electrical phenomena. Thus electricity, chemical affinity, heat, mechanical force, and the existence of matter were all to be referred now to the original pair of polar forces. He was willing to allow that a 'higher law, a higher principle of unity' ruled in organic than in inorganic nature, but stressed that this was not the same as saying that completely new forces (rather than new manifestations of known forces) were involved; 'the fundamental forces alone still constitute matter, whence arise all possible entities'.[178] This consideration gave him the opening he needed

---

[175] op. cit. (164), marginal note pp. 42–3: cited by K. Coburn, ed. *Inquiring Spirit*, pp. 249–50. London (1951).

[176] ibid., MS. note to pp. 86–9.

This seems to me the essential distinction, the others only consequences of this—viz. in the Alkalis the metallity (combustible quality) is uppermost, in Acids, the comburent, or antimetallic. But when I mediate on the term, Metallity, and that the metallic series form the axle tree on which all material Bodies circumvolve, that Carbon and Azote are the − and +, the N. and S. Poles of the Line (or shall I risk a yet bolder metaphor, and say, the solid cylinder?) of substantive Nature. I cannot suppress the suggestion, that the Qualitative Energies, the *inside*, of the metallic Bodies must be looked to, in order to discover the most proper character of Metallity,—and that one great purpose of the Noun adjectives Oxygen plus Chlorine plus Iodine, and Hydrogen is to *express* their qualities—by destroying or exhausting their quantitative and outside power of Cohesion or appropriative Attraction! That thus the *Contractive* and the *Dilative* restore the conditions, under which the Qualities can be called from potence into act.

Humphry Davy was engaged in some of his wilder speculations at about this time (see R. Siegried, *Chymia* **9**, 117 ff. (1964)); this may be more than coincidental, although I can find no external evidence to suggest that Coleridge influenced Davy at about this date.

[177] op. cit. (168), pp. 116–50.   [178] ibid., pp. 235–6.

for a discussion of the nature of matter and of force. In passing, he noted that it would be pointless to criticize the various systems. 'The atomistic system in general has long ago been subjected to sufficiently sharp criticism in Germany . . .' Opposite forces were the opposite manifestations of a higher unity (here was a partial answer to Coleridge's comments above), the single original force which pervaded all space. Bodies differed according to the intensity with which their forces filled space, through excess of one or the other force and its effects. It was only when the effective forms of forces had been investigated via their fundamental forces that a completely scientific chemistry would become possible.[179]

Oersted wrote about atoms, and his approach seems at times to have been wholly atomistic—but he insisted that his attitude was always that of the dynamist:

The fundamental difference between the atomist and the dynamist is always, that the former wants to construct the Whole from parts which are already complete, while the latter considers that the Whole, *with* its parts, *develops* as different forms of an original force. However, it is only at the boundaries of science that this difference becomes clear.[180]

On the level of mundane science, on the other hand, one could deal with nothing but the observational data provided by experiment. Coleridge was incensed at the suggestion that theories of matter were not very relevant to the prosecution of science; not only had all recent discoveries in chemistry been based on dynamical reasoning, but he knew that

since the year 1798 every experiment of importance had been distinctly pre-announced by the founders or restorers of the constructive or dynamic philosophy, in the only country where a man can exercise his understanding in the light of his *reason*, without being supposed to be *out of his senses*.[181]

The last sentence gives a good indication of the hostility that *Natur-philosophie* had to face outside Germany. As Thomson noted, however, within Germany the *Ansichten* 'has attracted great attention, and gained great *éclat*'. Outside Germany, the work was largely ignored,[182] and when Thomson finally acquired a copy of the French version he noted sadly that, although he found it impressive, it was still virtuallly unknown in England.[183] In 1815 Thomson had observed that some parts of the hypothesis struck him as 'whimsical and absurd', perhaps

---

[179] op. cit. (168), pp. 265 ff.   [180] ibid., p. 267.
[181] Letter to Lord Liverpool, 28 July 1817: quoted in Griggs, ed. *Collected Letters of S. T. Coleridge*, vol. 3, p. 760. 4 vols., Oxford (1956–9).
[182] Chevreul's approval was a rare exception; Harding, op. cit. (159), vol. 2, p. 291.
[183] *Ann. Phil.* **13**, 368 (1819).

because he had seen only abstracts and reviews; as for the metaphysical part, he completely failed to understand it. In his 1819 review, Thomson gave a fairly good summary of Oersted's ideas, with the exception of those on heat—which he found eminently satisfactory but incomprehensible.[184]

The *Ansichten* contained a well-developed theory of the identity of chemical and electrical forces, as the title of the French edition indicated. It also contained an early proposal for an investigation to provide experimental demonstration of the identity of electricity and magnetism, and their mutual convertibility.[185] In 1820 Oersted made his well-deserved discovery of electromagnetism,[186] which came as a long-anticipated and well-deserved confirmation of a philosophically certain conviction.[187] The usual international failure of communication set in; Davy first heard a confused account via Switzerland,[188] while in Paris, as Dulong wrote to Berzelius, when the news did arrive it was at first generally disbelieved. 'We thought it was another German dream.'[189]

In the late 1820s, Oersted became involved in a long correspondence with Weiss (whom he had met in 1802) on the nature of matter and of force. Oersted castigated the atomic theory as dead, sterile, and passive, and thought that if only one could understand the relations between the fundamental forces, one would have the key to the whole of natural philosophy. Nevertheless, he objected to calling his dynamical atomism 'anti-atomic'; this would make it purely negative in its implications. After pages of intense struggles with metaphysics, Oersted admitted the futility of such an approach to problems of science. 'Metaphysics has so far taught us nothing at all about the nature of bodies.' This was not to deny that metaphysics was a part of human investigation of the factors underlying experience, but it did mean that it could not function as judge in natural philosophy.[190]

Thus Oersted displayed the same careful balance and distinction between metaphysics and laboratory research as Davy did, but he placed his fulcrum far nearer to the former. Davy admired Oersted, whom he had met in 1823 and again in 1824, but, as we have seen, remarked disapprovingly that he was 'a little of a German metaphysician'.[191]

Davy's attitude must have been shared by many Englishmen, for

[184] ibid., 461–3. [185] See passage quoted in *Ann. Phil.* N.s. **2**, 321 (1821).

[186] *Phil. Mag.* **56**, 395 (1820); **57**, 40 (1821).

[187] R. C. Stauffer, *Isis* **44**, 307 (1953).

[188] H. Davy to J. Davy, 19 Oct. 1820: cited in J. Davy, ed. *Fragmentary Remains Literary and Scientific of Sir H. Davy*, p. 236. London (1858).

[189] Söderbaum, op. cit. (61), vol. 4, p. 17: letter of 2 Oct. 1820.

[190] Harding, op. cit. (159), vol. 1, pp. 281–331: letters of 1829.

[191] R. Instn MS. iv, 14i.

Oersted experienced considerable difficulty in finding a publisher for his philosophical dialogues. Sabine[192] had put him in touch with Longmans, who were not satisfied with the work, and Oersted then turned to Herschel for help. Herschel noted in 1844, 'Nothing came of this correspondence.[193] I was unable to be of any use in the matter';[194] and Oersted resigned himself to restricting publication to Denmark and Germany:

I know very well that many opinions, which find approbation in the germanic part of Europe, are not palatable to English readers. Your countrymen may in many respects be in the right, but I am in my mind assured that they go too far. I am far from approving the German metaphysics; but as I have studied those things now through more than fifty years, I have since long time been led to form a philosophy of my own, resulting from my studies both of experimental natural philosophy and metaphysics . . .[195]

In 1852, however, the collection was published in England as *The Soul in Nature*. The work, as Oersted wrote, was not about experimental researches, but instead presented 'the view of a natural philosopher upon subjects appertaining to the intellectual world'. Nevertheless, his wide-ranging dialogues inevitably illuminated some part at least of his science. His dynamical atomism is there, together with the concepts of duality and polarity in unity, and the emphasis upon forces. There is also, and significantly, an account of the sources of his philosophy of nature, which approximates closely to Faraday's theology of nature, and to various strands of Platonism in Christianity. This is why he could write to Herschel that he disapproved of German metaphysics, even though his philosophy often led to the same conclusions. '. . . the laws of nature are founded upon Reason . . . thoughts of Nature are also thoughts of God . . . all existing objects are active forces of nature, which represent to us a unity of thought . . .'[196] Indeed, the laws of nature are the thoughts of nature, which execute the ideas of nature. Reason is divine, either God's or in God's gift, and through its exercise man is able to perceive that all existence is 'a Dominion of Reason', and that there is an 'essential Unity of Intelligence throughout the Universe'.[197] In distinguishing Oersted from the Nature Philosophers, for whom as for Oersted Reason was in Nature, we should note that Oersted's Reason was the gift of a God transcending nature, while the Nature Philosophers considered that Nature and the Absolute comprehended one another.

[192] Harding, op. cit. (159), vol. 2, pp. 508–9: Oersted to Sabine, 6 Oct. 1848.
[193] ibid., vol. 2, pp. 401 ff.: letters of 1849.   [194] R. Soc. Herschel MS. xv, 171.
[195] Harding, op. cit. (159), vol. 2, p. 404: letter of 12 Oct. 1849.
[196] H. C. Oersted, *The Soul In Nature*, pp. 11, 19, 22. London (1852). See also D. M. Knight, *Studies in Romanticism* **6**, 82–8 (1967), for a discussion of this work in more general terms.   [197] Oersted, op. cit. (196), pp. 23 ff., 91–133.

Chemistry, in this context, was a more significant science than was generally realized, for its laws too were as much laws of Reason as were those of mathematics. Chemistry had advanced 'by the constantly combined influence of thought and experience . . . the newly discovered laws were always more and more freed from [the mist of error] . . . and stood forth as a necessity of reason'.[198] Experiment had shown the fundamental unity of heat, light, electricity, magnetism, and chemical affinity.

Added to this we already see numerous indications of a future, in which the chemical and mechanical laws of nature will be more intimately united.

In short, the natural laws of chemistry, as well as those of mechanics, are laws of Reason, and both are so intimately connected, that they must be viewed as a unity of Reason.[199]

The religious implications are evident, and Oersted had made them absolutely explicit in a lecture of 1814, where he described the cultivation of science as an exercise of religion.[200] Even earlier, in the winter of 1805–6, he had noted that science did not progress evenly;[201] this was related to the universal polarity in which he believed, according to which mental and spiritual life, in which he included the pursuit of science, consisted of a struggle of antagonistic forces, just as in the corporeal world. However, there was a Unity behind this duality.

While every thing in the great whole, down to the smallest part, varies between hate and love,—while the inquirer himself must share in this vicissitude,—while even his own human passions may be set in motion by the external impressions of nature, he may yet preserve security and repose amidst this vortex; indeed, I may venture to say happiness, if he only steadily fixes his eye on the firm unity, which no power on earth can destroy.[202]

The non-empirical features of Oersted's work are of real significance for his science and for the history of science. Oersted was convinced of the identity of chemical affinity and electrical forces. Faraday later provided experimental demonstration of this identity. Oersted discovered electromagnetism, which further advanced the nineteenth century's main endeavour in physical science, the correlation of forces, and was certainly the stimulus for Ampère's electrodynamical theory. All these influenced the development of ideas about affinity, but none of them arose exclusively from logically rigorous and scientifically impeccable origins. They were, instead, an integrated part of Oersted's whole outlook; if this outlook was not recognizably modern and scientific, it was eminently successful in furnishing him with theories that led to discoveries, and as such it deserves attention.

[198] ibid., p. 104. [199] ibid. [200] ibid., pp. 134–42.
[201] ibid., p. 320. [202] ibid., p. 324.

5

BERZELIUS

Jöns Jacob Berzelius embarked on his medical studies at Uppsala in 1796; the course began with chemistry and physics. Six years later, at the age of twenty-three, he presented a brief thesis on the therapeutic effects of galvanic currents.[1] In the same year he published a review of 'galvanic' research in chemistry;[2] at the very beginning of his career, he combined the eighteenth century's interest in galvanotherapy with the newly fashionable and exciting province of electrochemistry.

These early investigations into the effects of electricity in organic and inorganic nature doubtless contributed to his opinion, expressed in his lectures, that the same laws held in both realms. He denied, in opposition to Bichat, that there was a special vital force; rather, the phenomena of 'vitalism' stemmed from the organization of matter, and of the mechanical and chemical forces acting upon it.[3] The same chemical affinities therefore held in organic and inorganic nature, and, although organization distinguished these, preventing the former from being reduced to the latter, the laws of chemistry were in themselves everywhere invariable. Organization, and the functioning of nervous processes, might well prevent one from following the detailed elaboration of chemical processes in the organic realm; but one should remember that such processes were governed by chemical laws, and that any limitations on their study were merely practical. Convincing arguments have been put forward that young Berzelius did in fact view the problem of vitalism in this manner;[4] this was to prove very significant for his subsequent work, especially for his approach to the practical and theoretical union of organic and inorganic chemistry.

[1] 'De electricitatis galvanicae apparatu cel. Volta excitae in coropora organica effectu', thesis, Uppsala (1802). All unacknowledged biographical information in this chapter is taken from J. Jorpes, *Jac. Berzelius*. Stockholm (1966).

[2] J. J. Berzelius, *Afhandling om Galvanismen* (1802).

[3] For Bichat's vitalism, see M. F. X. Bichat, *Recherches physiologiques sur la vie et la mort*, pp. 79 ff. 3rd edn, Paris (1805).

[4] B. J. Jørgensen, 'Berzelius und die Lebenskraft', *Centaurus* 10, 258–81 (1964). A different interpretation has been put forward by J. Jacques, in a paper that fails to recognize numerous ambiguities in the contemporary expression of views on vitalism (*Rev. Hist. Sci. Applic.* 3, 32 (1950)). A detailed analysis of these ambiguities has been carried out by J. Brooke, 'Wöhler's Urea, and its Vital Force—a Verdict from the Chemists', *Ambix* 15, 84–114 (1968).

Berzelius, having submitted his thesis, had the good fortune to collaborate with Hisinger, an accomplished and experienced chemist.[5] Together, they electrolysed many salts and some of their bases.[6] Vauquelin belatedly ranked their results as highly as Davy's discovery of the alkali metals. Davy, to avoid any diminution in his reputation, had to state that he had remained ignorant of the Swedish chemists' work, until he had completed his own principal writings. This, although not entirely true, was at least plausible—the English were generally unable to read German, and almost unable to read scientific publications in Swedish. Berzelius, hampered by such provincialism, later complained:

The chemists of England live in their own world to such an extent that they are hardly aware of what is being done in France, and they are entirely unfamiliar with what is being done in Germany, about which they are as unconcerned as if absolutely nothing were going on there . . . It is quite difficult to get anyone here to believe that we were before Davy in discovering the ability of the galvanic pile to split salts and inorganic compounds in general into their parts.[7]

Hisinger and Berzelius noted that in electrolysis any given substance always went to the same pole, and that substances attracted to the same pole had other properties in common; there seemed to be at least a qualitative correlation between the chemical and electrical natures of bodies.[8] Their researches suggested also that the extent of electrolysis was proportional to the strength of the affinities between the components of the electrolyte, while the absolute quantity of decomposition depended directly upon the quantity of electricity that had passed.

Hisinger and Berzelius assumed that metallic salts were first decomposed into acid and base, and that the base then yielded metal and oxygen. The acid was merely a solvent, taking no further part in the process. They sensibly refused to commit themselves to a more detailed explanation of electrolysis. 'However, it seems most natural to us to explain it by the attraction which electricity has for some substances, and its repulsion for others, although this explanation strikes us as far from adequate.'[9] The attractive and repulsive powers of electricity upon matter were already assumed in *Naturphilosophie*;[10] but Berzelius and Hisinger dissociated themselves from this school by noting that the

---

[5] 'The Place of Berzelius in the History of Chemistry', *K. Svenska VetenskAkad. Årsb.* 34 (1948).

[6] J. J. Berzelius and W. Hisinger, in *Neues allg. J. Chem.* **1**, 115–49 (1803) (reprinted in *Ann. Phys.* **27**, 270–304 (1807)).

[7] Cited by Jorpes, op. cit. (1), p. 99.   [8] op. cit. (6), 143.   [9] ibid., 148.

[10] As G. Eriksson (*Lychnos*, 1–37 (1965)) has pointed out, Romantic dynamism was both known and recognized as such at Uppsala while Berzelius studied there; presumably the same is true of Swedish universities in general at that period.

repulsion exerted by electricity was merely a smaller degree of affinity or attraction.[11] They thereby implicitly identified electrical attractions with elective affinities, though they did not bring out the significance of this relation. Davy, in his first Bakerian Lecture of 1806, stressed the analogies between chemical and electrical attraction, and demonstrated the generality of their coincidence. Berzelius was probably much influenced by these ideas. He modified his original views, and embarked upon a new series of electrochemical researches.[12]

Prior to their estrangement in 1813, both Davy and Berzelius speculated on the constitution of ammonia and nitrogen. Berzelius and Pontin had discovered a means of amalgamating ammonia, using a mercury cathode. Berzelius told Davy of this, and they both investigated the nature of ammonia amalgam.[13] Davy and Berzelius considered that the volume of hydrogen evolved indicated that ammonia, and by inference nitrogen, contained oxygen gas. Theories of combining proportions proved useful as analytical tools in this investigation, and aroused Berzelius's interest in Dalton's atomic theory.[14] Dalton's *New System of Chemical Philosophy*, volume 1, part 2, containing the application of his atomic theory, did not appear until 1810; meanwhile, Berzelius was accumulating experimental results on chemical combining proportions. These led him to the discovery of 'several laws of affinity, which will render Chemistry more worthy of the name of Science, and through which our results will be more sure and more uniform'. His laws of affinity were in fact less generalized versions of Dalton's law of multiple proportions, but their formulation did not indicate a complete acceptance of Daltonian atomism. For example, Berzelius wrote that one part of a substance A might combine with 1, $1\frac{1}{2}$, 2, 4, ... parts of substance B. These results, and the experiments leading to them, owed much to Richter's work on combining proportions.

Berzelius was still eager to see Dalton's book, and so, presumably, not prejudiced against atomism; but his researches at this stage were carefully described in terms of combining proportions. Just so did Davy carefully avoid even the mention of atomism when describing Richter's observations on equivalent neutralization; he complained to Berzelius that Dalton's atomic theory was founded upon 'ideas more ingenious than correct'.[15]

---

[11] op. cit. (6), 148 n.   [12] See C. A. Russell, *Ann. Sci.* **19**, 117 (1963), and below.
[13] Söderbaum, ed. *Jac. Berzelius Bref*, vol. 2, pp. 8, 9, 16 etc. (6 vols. + 3 suppl., Uppsala, 1912–32): letters of 1808 between Berzelius and Davy.
[14] In June 1811 Berzelius still only knew of Dalton's ideas through Murray's *System of Chemistry* (4 vols., Edinburgh, 1806–7). The supplement of 1809 contained an account of Dalton's views. See Söderbaum, op. cit. (13), vol. 2, p. 29: letter to Davy, 10 June, 1811.
[15] Söderbaum, op. cit. (13), vol. 2, p. 22: letter of 24 March 1811.

Berzelius now combined his work on electrolysis with an investigation of the fixed proportions of the constituents of inorganic bodies, forming the opinion that chemical differences between bodies arose because of electrical differences between their constituents.[16] Thus hydrogen sulphide was an acid gas because of the electronegative character of sulphur.[17] The heat and light evolved in chemical reaction were clearly electrical in origin, and, as Davy's researches had clearly demonstrated, 'all phenomena of combination or of separation must be electrochemical operations. But we still know nothing of the nature of electricity.' One could, however, arrange bodies in order and attach numerical values to their electrochemical relations without assuming anything about the nature of electricity. Berzelius intended that his electrochemical series should be so interpreted as not to commit him to any theory of electricity.

Experiment, however, had shown that electricity seemed to obey the same quantitative laws as ponderable matter in chemical combinations, 'which', Berzelius noted, 'could indeed have been concluded *a priori*'. The next step was easy to take—a body's ability to neutralize the electric state of another, with which it was combined, was identical with its force of affinity for that body. Berzelius pursued this train in his next paper, the 'Essay on chemical nomenclature',[18] where he reiterated his suspicion that electricity was identical with chemical affinity. 'The different relations of bodies to electricity will henceforth be the basis of all chemical systems . . . [This problem] will soon become the general object of our researches, and gives us reason to hope for a new dawn in chemical theory.' Berzelius had come insensibly to adopt the phraseology of atomism, and now, when he proposed a theory of electrolysis,[19] he did so in terms of binary atomic reactions governed by electricity; he assumed, as a working hypothesis, that the latter was composed of two imponderable fluids.

Corresponding to this theory of electrolysis was an electrochemical scale of bodies, where oxygen uniquely possessed an absolute charge, against which the charges of other bodies were measured. Organic compounds, unlike inorganic ones, could be ternary or quaternary. They were formed by the influence of the nervous system, which operated electrically and maintained the elements in modified electrochemical states; this modification relaxed as soon as the substances were isolated from the organized and organizing system, when the elements immediately reverted to their 'normal' states—hence putrefaction, fermentation, and other processes involving the decomposition of organic

[16] *Ann. Chim.* **79**, 233 ff. (1811).
[17] Berzelius briefly adopted the opposite sign convention for electric charges.
[18] *J. Phys.* **73**, 252 ff. (1811).   [19] ibid., 277–8 n.; see also note (12).

A.M—L.

compounds free from organizing influences to simpler inorganic substances.[20]

Thus, not only was electricity the key to chemical affinity, it was also, through the organizing power of the nervous system, the reason for the differences between organic and inorganic compounds. These differences were most easily seen in the laboratory in complexity of chemical constitution. At times 'complexity' seemed a hopeless under-statement, for accurate organic analyses were extremely difficult to perform. With skill, however, one could achieve good results. Berzelius wrote to Gay Lussac in the autumn of 1811 congratulating him on his analyses of three carbohydrates.

It would be difficult to do two experiments on the same substance which would agree better with one another than the analyses of these different substances. What is the reason for their different chemical characters? . . . How shall we ever arrive at clear ideas about this problem without electro-chemical science?[21]

Difference in electrochemical states, rather than in arrangement, was the first thing that occurred to Berzelius in considering these substances; the electrical natures of individual atoms would, he believed, account for molecular properties.

In 1812, Berzelius visited London, where he met Wollaston and Davy.[22] Dalton at last presented him with a copy of his book,[23] and Wollaston and Berzelius had long discussions on the atomic theory in relation to definite proportions. Dalton had attacked Gay Lussac's law that gases combined in simple proportions by volume, since there seemed no mechanical reason for it.[24] Berzelius, on the other hand, thought that this law was the finest proof of the atomistic theory;[25] his acceptance of it at this date also appears from the second part of his *Textbook*.[26] Note, however, that atomism did not have to be Daltonian.

Davy met Berzelius in London a few days prior to leaving for Scot-land. His *Elements of Chemical Philosophy* appeared after his departure, so that he and Berzelius were not able to discuss it together. Berzelius

[20] *J. Phys.* **73**, 260–1; see also Russell, *Ann. Sci.* **19**, 136–40 (1963).

[21] Söderbaum, op. cit. (13), vol. 7, p. 115: letter of 25 Sept. 1811.

[22] ibid., vol. 1, p. 41: letter to Berthollet, [Oct.] 1812; see also Jorpes, op. cit. (1), pp. 59 ff.

[23] Söderbaum, op. cit. (13), vol. 2, p. 32: letter to Davy, summer 1812.

[24] H. E. Roscoe and A. Harden, *A New View of the Origin of Dalton's Atomic Theory*, p. 159 (London (1896)): letter from Dalton to Berzelius 20 Sept. 1812. For Dalton, geometrical and mechanical factors were paramount in chemical combination: ideas about the force of chemical affinity would have seemed gratuitous. See, for example, Instn elect. Engrs MS. 6 May 1834, note from lecture, where he discussed combination without even alluding to affinity.

[25] Roscoe and Harden, op. cit. (24), p. 162: letter of Berzelius to Dalton, 16 Oct. 1812.

[26] Eriksson, op. cit. (10), 36.

was disappointed when he read it. 'I am altogether unable to recognise in the author of this work the discoverer of the alkalis and of the earths. He spends far too much time on hypotheses whose very probability strikes me as suspect.'[27] Berzelius, more cautious and more widely familiar with chemical literature than Davy, was conscious of the latter's errors and sceptical about his hypotheses. For example, Davy's explanation of heat as motion seemed both vague and premature. If as much was known about caloric as he implied, then an understanding of its effects upon chemical affinity would be within reach; but this was clearly not the case.[28]

Berthollet's arguments about mass action also gave them grounds for disagreement.[29] Davy thought that mass action was incompatible with fixed proportions; and so it was, given his concepts of chemical combination and matter. Berzelius, no less attached to the empirical facts of definite proportions, nevertheless thought that the two principles could be reconciled. Although he distinguished between solution and chemical combination, he invoked forms of chemical affinity to account for the stability of both. Chemical combination was generally exothermic. Dissolution was frequently endothermic; it depended upon a specific affinity between solvent and solute molecules, but nothing precise was known about the nature of this force.[30] In solution, the equilibration of affinities produced a symmetrical disposition of solvent and solute molecules. Consider a system with two solute species, A and B, where A has a greater affinity than B for a third species; then the affinity of a smaller number of solute A molecules will be balanced, at equilibrium, by the affinity of a larger number of solute B molecules. Thus, for example, in an aqueous solution of copper nitrate, to which sulphuric acid has been added, 'free atoms of both acids will arrange themselves around those which remain in combination, and will prevent, by means of their opposed and counter-balanced forces, their mutual combination with the cupric oxide.'[31] Berzelius considered that Berthollet's theory of mass action followed necessarily from the assumption of a system where all proximate affinities were in equilibrium. Berzelius continued to accept the laws of combining proportions, both as facts and as logical consequences of the atomic theory, and he showed that mass action was compatible with atomism. Compatibility, however, does not confer necessity.

Berzelius did appeal to geometrical factors to indicate the saturation

[27] Söderbaum, op. cit. (13), vol. 2, p. 41.
[28] ibid.
[29] See Chapter 2, notes (143)–(147).
[30] H. Das, Over de historische ontwikkeling van het begrip 'moleculverbinding', p. 16 Thesis, Amsterdam (1962).
[31] J. J. Berzelius, *Traité de chimie*, trans. Jourdain and Esslinger, vol. 4, pp. 576–7. 1st Swedish edn (*Lärbok i kemien*), Stockholm (1818), this edn, 8 vols., Paris (1829–33).

point in binary combinations and to suggest the equilibrium configurations of a solution; but he obscured his argument by insisting that solution was not combination, since it involved a different affinity. Davy was wrong to argue that, given mass action, there could be no fixed proportions; but if we assume that the same forces operate in solution and in solid combinations, and, like Berthollet, take solutions as the paradigm, Davy's argument becomes valid. Davy and Berzelius differed both in their choice of paradigm and in their assumptions about the forces operating. Without a clear statement of these presuppositions, discussion was unprofitable. Accordingly, Davy did not even attempt to discuss mass action in his chilly answer to Berzelius's criticisms. He merely thanked his mentor 'for your good intentions in criticising my work. It is sufficiently popular, but I shall endeavour to make it more so.'[32] Preparations for a second edition were wasted;[33] the first volume of the first edition was also the last.

Berzelius failed to recognize the lacunae in his synthesis of mass action with definite combining proportions; in 1813 he published an account demonstrating the relation between Berthollet's theory of affinities and the laws of chemical proportion, but could show only 'that the principles of Berthollet's theory are not inconsistent with the laws of chemical proportions'.[34] His model of chemical combination required that those parts of bodies that combined should do so in definite proportions; any excess of the various substances would remain in an equilibrium determined by their antagonistic forces. Atomism and mass action were merely assumed successively, rather than shown to be necessary consequences of a unified theory of matter and force.

Berzelius realized that if only this theory, embracing the cause of definite combining proportions, could be elucidated, then the 'principal basis of chemical theory' would stand revealed. There were, undoubtedly, attractions in Dalton's plausible and simple mechanical hypothesis, but for all its clarity 'it is connected with great difficulties'.

These unspecified difficulties were surely aggravated by Dalton's scant attention to the problem of chemical affinity. This was a matter of prime concern to Berzelius, who therefore stated, for the first time, his own notion of the corpuscular theory, including the 'mechanism' of chemical combination.[35] The result was a model that gave a more detailed account of chemical affinity than any previous hypothesis. The degree of specification was such as to make later modification difficult, and Berzelius's adoption of this inadequately open-ended model

---

[32] Söderbaum, op. cit. (13), vol. 2: letter of 4 Aug. 1813.
[33] A copy, annotated by both John and Humphry Davy in preparation for the second edition (never realized), is in the Royal Institution.
[34] *Ann. Phil.* **2**, 443 ff. (1813).   [35] ibid., 446 ff.

accounts for some inflexibility in his subsequent career.[36] Berzelius envisaged all atoms as being mechanically indivisible spheres of equal radius. The atoms touched when they combined with one another. The difference between homogeneous and heterogeneous aggregates was that in the latter 'an electric discharge takes place of the specific polarity of the heterogeneous atoms, which cannot take place between homogeneous atoms'.

Dalton, who seems to have regarded the entire preserve of corpuscular chemistry as his own, was moved to protest at Berzelius's unorthodox atomism. Every point instanced above was dismissed by Dalton as being no part of his doctrine; in particular, 'What Dr. Berzelius says of the *electric polarity* of atoms . . . makes no necessary part of the atomic theory such as I maintain.'[37] Naturally not, for Dalton's theory was essentially descriptive in intention, while Berzelius aimed at the explanation of affinity, in terms of its *modus operandi*.[38]

Berzelius went on to develop the dualist system implicit in his theory of electrically polar atoms. He believed that inorganic substances were made up of biatomic molecules AB, or of bimolecular aggregates AB.CD, as in the case of salts. Organic substances differed from inorganic ones because their constituent molecules could contain more than two elementary substances. Geometrical considerations suggested that an atom A could be surrounded by twelve near neighbours B, since all atoms had the same volume.

If, on the other hand, we pay attention to the electric polarity of the atoms, an atom of A cannot combine with more than 9 atoms of B, if the atom

[36] cf. Sir Harold Hartley, *Humphry Davy*, p. 54. London (1966).
[37] *Ann. Phil.* **3**, 174 ff. (1814).
[38] *Ann. Phil.* **5**, 122 ff. (1815):

It has given me great pain to think that the respectable Dalton has taken my ideas on the corpuscular theory as a criticism on his . . . I neither meant to give the opinions of Dalton, nor a correction of them . . . Mr. Dalton states that the electro-chemical polarity of the atoms makes no necessary part of the atomic theory, such as he maintains; nor did I ever mean to convey any such idea to the reader. For my own part, in considering a corpuscular theory of chemistry, I conceived that it should constitute the fundamental theory of the science; and instead of being occupied with a part of the phenomena, ought to embrace the whole. But when we treat atoms in a chemical theory, we ought to endeavour to find out the cause of the affinity of those atoms. We ought to endeavour to combine researches respecting the cause why atoms combine with researches into the cause why they combine only in certain proportions. I do not consider the conjectures which I hazarded on the electro-chemical polarity of the atoms as of much importance. I scarcely consider them in any other light than as an ideal speculation deriving some little probability from what we know of the chemical effects of electricity. Yet the ideas on the relation of the atoms to their electro-chemical properties, ought in my opinion to constitute an essential part of the corpuscular theory of chemistry, such as I view it; because I consider it as the duty of a man of science to endeavour to reach the first principles of the science, even though it should be actually impossible to attain it.

A + 9B preserve any part of the electrical polarity originally belonging to A: for example, *oxymuriatic acid*, which is a compound of 1 atom of muriatic radicle with 8 atoms of oxygen, still preserves a part of the original polarity of the radicle, by means of which it reacts.[39]

Berzelius's account of the corpuscular theory and the terms he used show clearly that he would have liked to adhere to some form of corpuscular chemistry. Certain difficulties, however, had to be overcome before chemical atomism could be unreservedly adopted. For example, the molecular formula of some metallic oxides indicated the presence of half atoms of oxygen. One might avoid this problem by assuming that future knowledge of the full range of combinations would lead to different formulae without half atoms. Organic chemistry, as ever, presented difficulties that were less easy to surmount. Granted that elementary atoms would not combine with more than twelve other such atoms, how could one make sense of the composition of organic bodies? Oxalic acid, for example, according to Berzelius's own analysis, 'consists of one atom of hydrogen combined with forty-five other atoms'—and even a theory of compound radicals could not reconcile this with the dictates of geometry. Berzelius, therefore, while sure in his own mind that some sort of corpuscular theory was needed, refrained from an open commitment, and stated a preference for explanations in terms of volumes.[40]

His avowed caution in this respect was merely appended as a qualifying afterthought to an atomistic account. In the same year he published a much fuller account of the relations between chemical affinity and electricity, writing confidently about polar atoms and electric fluids.[41] The experiments of the previous decade had done more than demonstrate the influence of electricity upon affinity; they had proved 'that every chemical process is always an electrical process, and in general, that it is not possible for an affinity to manifest itself without the co-operation of the electricities'.

Berzelius supposed that atoms were accompanied by electricities;[42] when the former combined, so did the latter, their discharge appearing as heat. Thus the distribution of electrical fluids preceded chemical affinity, and was the necessary precursor of chemical combination. Berzelius repeated his classification of bodies according to their behaviour in electrolysis, with oxygen as the uniquely and absolutely negative body. Then, in a footnote that appeared in Nicholson's *Journal* but was omitted from the French version of the paper, he expounded his conjectural theory of affinity. The following passages, taken from that

[39] *Ann. Phil.* **2**, 447 (1813).   [40] ibid., 450.
[41] *Ann. Chim. Phys.* **86**, 146–74 (1813); **87**, 113–52; *Nicholson's J.* **34**, 142, 153, 240, 313 (1813); **35**, 38, 118, 159; **36**, 129.   [42] cf. note (20) above.

footnote, give a compact statement of his ideas. The passages are given *in extenso*, because, although the overall system is clear, details are problematic, rendering summary suspect.

Berzelius began by indicating the different polarities that atoms possessed. First, there was a general electrical polarity:

upon the intensity of which the force of their affinity depends. In this case, the chemical affinity becomes identified with electricity, or rather the electric polarity. In order to explain the different electro-chemical characters, we must add to the general polarity a kind of specific unipolarity,[43] by means of which one of the poles contains more of the $+E$, or the $-E$, than the opposite electricity in the other pole is capable of saturating. A body of which the positive pole predominates, i.e., which contains an excess of positive electricity, constitutes an electropositive body, and vice-versa. Many bodies require an elevation of temperature to enable them to act upon each other. It appears, therefore, that heat possesses the property of augmenting the polarity of these bodies, and that the difference in activity of the same affinity at different temperatures, appears to depend upon the same cause, in like manner as the force with which a combination preserves its existence, appears to depend on the intensity of the electric polarity when this is at its maximum, or rather the intensity of that polarity at the moment the combination is made. This circumstance explains why the phosphoric acid is decomposable by charcoal at an elevated temperature, although phosphorus decomposes the air of the atmosphere at a temperature at which charcoal has no influence upon that fluid.

He went on to distinguish homogeneous adhesion from heterogeneous combination, using his model of electrically polarized atoms to explain the difference between the two attractions:

In the theory of atoms, there is some difficulty in conceiving the difference between the juxta-position of homogeneous particles, separable by mechanical means, and that of heterogeneous particles, which produce a new particle, very seldom decomposable by means merely mechanical. The hypothesis of polarised atoms assists us upon this occasion. The cohesion of homogeneous particles may be compared to the juxta-position which we observe in the electrophore between the opposite electricities of the metallic plate and the resinous surface. Contact keeps them in a state of charge or neutralisation; which, in fact, is simply juxta-position, and is destroyed when the surfaces are separated, and each appears again in possession of its original electric state. When heterogeneous atoms combine (whether the combination do consist simply in juxta-position, or, which is more difficult to comprehend, in a partial or total penetration) they appear to adjust or dispose themselves so as to touch with the opposite poles; of which the electricities produce a discharge which causes the phenomenon of elevation of temperature, almost constantly apparent at the time of any chemical combination, and the

[43] Erman, *Ann. Chim. Phys.* **61,** 115 (1807).

particles remain combined until their discharged poles are, by some means or other, restored to their former electric state.

Berzelius, having indicated the electrical mode of chemical combination, considered next the relation between the electrical polarities and elective affinities of atoms. His conclusions enabled him to explain why some physical states were particularly favourable to chemical combination, and also implied a mechanism for electrolysis:

As we know, from fact and experience, that bodies of the same electrochemical class (i.e. to say, bodies in which we conceive the same pole predominates) can combine, it appears, that the force of affinity depends rather on the intensity of the general polarity, than the specific unipolarity; and from this reason it may be, that sulphur has more affinity with oxigen, than gold or platina has, although sulphur has the same unipolarity as oxigen, and these metals have an opposite unipolarity to that of oxigen.

It is clear, that when two bodies, in which the same pole predominates, combine together, the new particle must possess their unipolar forces concentrated in one of its poles, and must, consequently, have electrochemical properties more intense; and this is a good reason why sulphur and oxigen produce the strongest acid. On the contrary, when particles possessing an opposite polarity unite, the polarity of one of the particles most frequently predominates, for example, in potash, and in most of the metallic oxides, the predominating pole of the metal also predominates in the compound. In some instances, the production is a neutral compound, in which neither of the poles predominate, such as the superoxides: in other instances, the pole of the metal predominates in one degree of oxidation, and that of the oxigen in another.

The combination of polarized atoms requires a motion to turn the opposite poles to each other; and to this circumstance is owing the facility with which combination takes place when one of the two bodies is in the liquid state, or where both are in that state; and the extreme difficulty, or nearly impossibility, of effecting a union between bodies, both of which are solid. And again, since each polarized particle must have an electric atmosphere, and as this atmosphere is the predisposing cause of combination, as we have seen, it follows, that the particles cannot act but at certain distances, proportional to the intensity of their polarity; and hence it is that bodies which have affinity for each other, always combine nearly on the instant when mixed in the liquid state, but less easily in the gaseous state, and the union ceases to be possible under a certain degree of dilatation of the gases, as we know by the experiments of Grothuss, that a mixture of oxygen and hydrogen in due proportions, when rarified to a certain degree, cannot be set on fire at any temperature whatever.

The chemical action effected by the discharge of the pile, consists in the particles in a combination being repolarized. In a combination of particles having the same unipolarity, the pile merely restores, by the decomposition, the general polarity, because their specific unipolarity was not changed by their union; but in combinations of opposite unipolarity, it likewise restores the specific unipolarity of the elements. May we conclude, that in the first case, the general repolarization takes place in the same manner as the load-stone gives magnetism to a small particle of steel, and that in the second, the pile contributes, by its own specific energies, to restore the predominating poles.[44]

Berzelius's theory of affinity differed from Davy's in three major respects, stating as it did that each atom had unlike and specific polar charges, which were the seat of affinity, that combination involved neutralization of charge, and that, in electrolysis, reacting components first combined, losing their charges by neutralization, and then regained their charges from the pile.[45] At a different level of interpretation, Berzelius regarded affinity as arising from electrical relations, while Davy believed that affinity and electricity were different manifestations of the same cause; their viewpoints both arose naturally from their own notions of the nature of electricity.[46] Again, Berzelius's theory was a dualistic theory, in contrast with Davy's, which, for all that it relied upon polar forces, was nevertheless fundamentally a unitary theory, allocating quite as much importance to the overall arrangement of powers and elements as it did to the specific electrochemical character of the individual elements. Berzelius could happily claim that, since acids and bases were both binary compounds of oxygen with a radical, acidic and basic character were functions of the electrochemical charac-ter of the radicals alone;[47] to Davy, the modes of combination, including arrangement, were also relevant. Faraday, strongly opposed to the concept of unipolarity, differed even more strongly from Berzelius. Faraday's quantitative laws of electrolysis, correctly understood, also argued decisively against Berzelius's views; if a given amount of elec-tricity liberated equivalent quantities of different substances, then the electrical charges of atoms could not be functions of their chemical activity.

Berzelius concluded his exposition by remarking that the whole theory, as he had proposed it, was still no more than a hypothesis in need of experimental confirmation, or, perhaps, of disproof. Such a beautiful generalization of facts would have been hard to relinquish,

[44] *Nicholson's J.* **34,** 154–6 n. Perhaps the note was omitted from the French version because Berzelius was then still relatively unknown in France, and, as A. Marcet wrote to him about Parisian chemists (Söderbaum, op. cit. (13), vol. 3, p. 127: letter of Nov. 1815): 'J'y ai trouvé des gens à grand nom qui disent que vos écrits sont pleins de vues hypothétiques et que vous vous amusez à symetriser la chimie en la soumettant à des lois imaginaires. Je vous repète qu'on ne vous connaît point à Paris.'
[45] cf. J. Ellowitz, M.Sc. thesis, London University (1927), p. 45.
[46] *Ann. Chim. Phys.* **86,** 168–71 (1813). The *imponderabilia* are probably material.
[47] ibid., 166–8.

and there is little doubt that Berzelius, in spite of his disclaimer to Dalton, was serious about his theory.[48, 49]

His acceptance of a theory of polar atoms, at least as a leading element in a working and guiding hypothesis, emerges from his attempt to adapt his theory of combining proportions to the realm of organic chemistry.[50] Berzelius viewed inorganic, and organic chemistry as twin branches of a single subject.[51] Thus in 1814 he wrote, 'It is evident that the existence of definite proportions in inorganic bodies leads to the conclusion that they exist also in organic bodies.'[52] Berzelius recognized that organic atoms were more complex than inorganic ones, and that the electrical states of their elements were variously modified; all substances, however, obeyed the same laws. The assumption of 'complex atoms' implied that the mechanical structure of organic substances would differ from that of inorganic ones. Knowledge of these structures could be embodied in models like Wollaston's[53]—how fully Berzelius now accepted the corpuscular doctrine—and might prove useful in correcting poor analyses, as these 'might probably indicate a number of elementary atoms, incapable of forming any regular figure whatever'.[54] Berzelius was concerned with the interactions between chemistry, mineralogy, and crystallography. He noted that there were simple atomic differences between certain organic substances, for example

$$
\begin{aligned}
\text{'sugar of milk':} \quad & 8O + 10C + 16H \\
\text{and 'mucous acid':} \quad & 8O + 6C + 10H, \\
\text{differing by:} \quad & 4C + 6H.
\end{aligned}
$$

Such instances were suggestive of 'the idea that organic atoms have a certain mechanical structure, which enables us to deprive some of them of certain elementary atoms without altering the whole very much'.[55]

This idea may well have contributed to the formation of Laurent's nucleus theory and of his crystallographic and structural approach to chemical types. Laurent looked to Berzelius for support for his ideas against those of Dumas, and was puzzled when he did not receive it.[56] Berzelius, however, in spite of his temporary inclination towards a structural approach to chemistry, continued to award primacy to the electrical polarity of atoms, with structure as a consequence of the

[48] See note (38) above.
[49] cf. *Ann. Phil.* **5,** 98 n. (1815): 'I acknowledge . . . that . . . I do not lay much stress on that argument [i.e. his reservations of 1813] against the corpuscular doctrine.'
[50] *Ann. Phil.* **4,** 323 (1814); ibid., p. 403. Berzelius tells us that he undertook his organic analyses 'in order to determine how far the laws which I had ascertained in inorganic nature would apply also to organic bodies'.
[51] cf. J. H. Brooke, 'There is but one Chemistry': paper read before the Society for the Study of Alchemy and Early Chemistry, 7 Dec. 1968.
[52] See note (50) above.   [53] *Ann. Phil.* **5,** 274 (1815).   [54] ibid.
[55] ibid.   [56] See Chapter 6 below.

attempts of individual elemental polarities to achieve their maximum mutual saturation. Metastable bodies, such as fulminating silver, contained elements whose polar strain was at a maximum; discharge, and consequent decomposition, occurred when the atomic polarities could rearrange themselves so as to yield a less excited state. Decomposition by explosion therefore always produced new compound bodies.[57, 58] Berzelius ignored structural arguments when discussing the formation of minerals also, and emphasized instead the agency of more or less weak electrochemical affinities.[59]

There was no compelling reason to alter this approach until 1819, when Mitscherlich published his discovery of isomorphism, significant for chemistry and crystallography alike. Mitscherlich's paper established numerous cases of isomorphism, engendering his extravagant hope that a study of crystalline structure would prove as effective as chemical analysis in revealing chemical composition.[60] He generalized the phenomena in his second paper on the topic, claiming that 'an equal number of atoms, combined in the same way, produce the same crystal forms, and the crystal form does not depend on the nature of the atoms, but only on their number and mode of combination'.[61] The phrase 'mode of combination' might seem to be susceptible of a variety of interpretations, and to include the influence of chemical affinity. Berzelius, however, offered an unambiguous description of isomorphism; since the work had been carried out in his own laboratory, he presumably knew what Mitscherlich meant. Berzelius wrote to Haüy that isomorphic substances possessed identical crystalline form, but differed in chemical constitution, 'without this variation being determined by chemical affinity, and without it conforming to the law of chemical proportions; it derives from the simple circumstance that these different substances can form integrant parts of the same crystalline form'.[62] Haüy, whose

[57] *Ann. Phil.* **7**, 431 (1816).

[58] R. Harrington, in *An Elucidation and Extension of the Harringtonian System of Chemistry, explaining all the Phenomena, without one Single Anomaly* (London, 1818), besides wishing that Davy, that self-styled Hercules, could be sent to the moon with all other modern chemists, continued his attack on nineteenth-century chemistry with great relish, and hoped that the eighteenth-century German schools would be revived and would 'hurl these empyrics to the dust' (p. 52). Berzelius's theory of metastability inevitably came under attack from this eccentric quarter (p. 61):

Berzelius says these combustions and explosions are produced by electricity from their attractions: then how comes separation of these attractions to produce explosions and fire,—you gross ignoramus? In short, they are pushed to the last extremity, and invent the most gross puerilities and absurdities upon the ignorant; expecting their names will give them currency.—These star chambers of science.

[59] *Edinb. phil. J.* **1**, 63, 243 (1819).

[60] *Abh. k. Akad. Wiss. Berlin* 427 (1819): trans. in *Ann. Chim. Phys.* **14**, 172 (1820).

[61] *Q. J. Sci. Arts* **14**, 198, 415 (1823).

[62] Söderbaum, op. cit. (13), vol. 7, p. 162, letter of April 1820.

crystallography was incompatible with this 'simple circumstance', naturally rejected the doctrine of isomorphism.[63] Berzelius, faced with Haüy's dogmatism, hesitated; but when he had seen for himself the accuracy of Mitscherlich's results, he became convinced, and welcomed isomorphism as a new link between chemistry and crystallography.[64] Such an approach would help determine the modes of combination of elementary atoms in crystals; the theory of isomorphism implied that these modes might be independent of forces, so that alterations were needed in the system of 'chemical mineralogy'. Berzelius, in his treatise on mineralogy, had arranged minerals according to the electrochemical character of the positive constituents;[65] isomorphous replacement between these constituents invalidated his classification, and in 1824 he proposed a new system based on the electrical order of the negative constituents.[66] Thus, although isomorphism at first impelled Berzelius to regard structure as being more important than he had previously allowed, he was clearly unwilling to let it displace electrochemical dualism as the basis of his system.[67]

Had Berzelius wished to maintain that crystalline structure was independent of chemical reactivity, he would surely have urged this as an argument against the theory of types. His failure to do so strengthens the argument that he came to regard structure and reactivity as being intimately connected. He believed that different arrangements of the same atoms would give rise to different chemical forms, and that experiment 'even seems to show that the difference in forms is accompanied by a modification in chemical properties'.[68] The arrangement of atoms, however, both in solution[69] and in crystals,[70] was determined by the equilibration of atomic forces, so that atomic properties were primary and arrangements secondary.

Mitscherlich came to realize later that the rigidity and generality of his first statement of the law of isomorphism were unsatisfactory. Only

[63] Söderbaum, op. cit. (13), vol. 7, p. 162: letter of 18 Oct. 1821.
[64] ibid., vol. 7, p. 124: Berzelius to L. J. Gay-Lussac, 25 Sept. 1820.
[65] J. J. Berzelius, *An Attempt to establish a pure scientific system of mineralogy by the application of the Electro-Chemical Theory and the Chemical Proportions*, trans. John Black (1814).
[66] *Ann. Phil.* N.S. **11**, 381, 422 (1826).
[67] He did, however, briefly attach much greater importance to mechanical structure than he had done before (or was to do again): in *Ann. Phil.* N.S. **4**, 356 (1822) he attributes instability either to weak affinities, *or* to unstable mechanical construction. Except at this date, the latter factor would scarcely have weighed with him.
[68] J. J. Berzelius, *Essai sur la théorie des proportions chimques, et sur l'influence chimique de l'électricité*, trans. from Swedish, pp. 25 ff. 2nd edn, Paris (1835).
[69] ibid., p. 65. cf. Berzelius, op. cit. (30), vol. 4, pp. 585–6.
[70] In the first edition of his *Lehrbuch der Chemie* (trans. from 2nd Swedish edn of *Lärbok i kemien*, this edn, 1825–31), Berzelius attributed crystalline structure to atomic polarities, and not vice versa. (Vol. 3, pp. 77 ff. (1st edn, Dresden, 1827) implies the latter.)

certain elements could replace one another. There was also some slight variation in the angles and forms of isomorphous crystals. Mitscherlich accounted for these anomalies by supposing that the nature of the chemical affinity between base and acid sometimes affected the dimensions of the crystals they formed.[71] He had foreseen that such an effect might exist as early as 1821,[72] and his later ideas on chemical combination were a synthesis of electrochemical dualism and structural considerations, both factors being essential and of comparable importance.[73]

Although Berzelius held his electrochemical theory of affinity and the concomitant version of the atomic theory with increasing confidence as evidence accumulated, he was not as rigidly committed to dualism as is generally supposed. In 1834, for example, Berzelius wrote to R. Hare that potassium sulphate could be split in different ways, such as $KO . SO^3$ or $K . SO^4$;[74] the former representation had hitherto been generally adopted, but as chemical theory developed, it became clear that both forms should receive equal consideration. This suggests that Berzelius fully accepted the underlying theory, but regarded the details of its application as remaining open. His formal representations of molecular constitution must therefore be regarded as merely conventional. The atomic theory was fundamental to his work; but he was cautious, and was therefore delighted when Wollaston seemed to offer experimental support for atomism.[75] Berzelius considered that 'it is very important that the corpuscular theory can be proved and considered as something more than hypothesis'.[76]

When, in 1836, Berzelius came to discuss and christen catalytic phenomena, he did so with his usual cautious glance towards electrochemical dualism.[77] The catalytic force occurred widely in both organic and inorganic chemistry, and, for lack of adequate knowledge, he called it a new force. This was not intended to imply that it was independent of the electrochemical relations of substances: 'On the contrary, I can only suspect that it is a special form of the action of the latter [electrochemical affinity].'

Until the end of his life, Berzelius adhered to almost the same electrochemical theory of affinity, with its stress upon the role played by the nature of individual atoms. He sought to apply it, at least in principle, to every problem that arose in chemistry. For example, when the allotropy of certain elements was demonstrated, he suggested that it was

[71] *Rep. Br. Ass. Advmt Sci.* p. 230 (1832).
[72] *K. Svenska VetenskAkad. Handl.* 48 (1821), quoted in trans. in op. cit. (71).
[73] *Ann. Chim. Phys.* (3) **4**, 67–76 (1842); **7**, 5 (1843).
[74] Söderbaum, op. cit. (13), vol. 7, pp. 141–2: letter of 23 Sept. 1834.
[75] Wollaston, *Phil. Trans. R. Soc.* **112**, 89–98 (1822).
[76] Söderbaum, op. cit. (13), vol. 3, p. 233: letter to A. Marcet, 14 March 1822.
[77] *Jahresber. Fortschr. Chem.* **15**, 237 ff. (1836).

due to the existence of different atomic states, and from this went on to propose these different states as the explanation of isomerism.[78] The nature of atoms, and not their position, was, as Dumas recognized,[79] the fundamental factor in Berzelius's system.

Berzelius did not assume an identity between organic and inorganic chemistry, but he did use his greater knowledge of the latter to provide him with an approach to the study of the former.

Although the composition of organic bodies, at the first glance, appears to be altogether different from that of the inorganic, yet what we know of the latter is the only unerring guide which we possess to enable us to form an opinion upon the former. In exploring the unknown, our only safe plan is to support ourselves upon the known. This must also be the right way here, and what we already know concerning the laws of combination which regulate inorganic nature, must be taken as a guide in judging the modes of combination in organic nature. Every other mode of procedure allows full scope to the fancy, which, varying only with the inventive faculties of the individual, is always ready to build new castles in the air. It is thus that numberless different views are set up and varied in all kinds of ways, no one following the same rule as the other, they cross and contradict one another in all directions; and this will continue, until we are all agreed by what rule the formation of our judgement should be guided. I therefore repeat, *that the application of that which is already or can hereafter be known concerning the laws of combination in inorganic nature, is the only guide to our researches concerning their mode of combination in organic nature; that by this means alone we can hope to arrive at a correct and unanimous opinion concerning the constitution of those bodies which occur in nature or which arise from the action of chemical agents upon them.*[80]

A major cause of the debate from the late 1830s until the early 1850s between dualists and those who accepted some form of the theory of types was the latter's attempted inversion of this regulative principle.[81] Once this inversion had been accomplished, further discoveries led once again to a reversal of direction. It was not until the analogy had proved experimentally fruitful when applied in both directions that chemistry really became a single subject. Superficially, the debate was little concerned with chemical affinity—certainly not as Berzelius understood it. But affinity, in the broadest sense, is an essential component of all but positivistic systems of chemistry;[82] the theory of types, being far from

[78] *Scient. Mem.* **4**, 240–52 (1846).

[79] *C. r. Acad. Sci., Paris* **10**, 171 (1840).

[80] *Scient. Mem.* **4**, 662–3 (1846).

[81] For a detailed discussion of the analogical arguments, see J. H. Brooke's doctoral thesis on the role of analogies in the development of organic chemistry (Cambridge University, 1968).

[82] Comte himself condemned affinity as metaphysical—see *The Positive Philosophy*, trans. H. Martineau, vol. 1, p. 300. 2 vols., London (1853).

positivistic, had important implications for affinity theory. The next
chapter will indicate what concepts replaced those of Berzelius's, and
how the two paradigms finally yielded to a third that was both more
and less than a synthesis of its two predecessors. The whole to-and-fro
movement will be considered in terms of the development of the concept
of chemical affinity.

6

CONTRIBUTIONS FROM ORGANIC CHEMISTRY

In the first third of the nineteenth century, inorganic substances alone seemed to obey more or less simple laws. Organic chemistry resembled nothing so much as cookery by trial and error. In spite of all the determined spade-work of eighteenth-century and even earlier chemists, preoccupied with problems of preformation and classification within the organic realm, organic chemistry as a coherent science dates only from around the 1830s. Its provenance, its legitimacy, and its scope were all attacked, and the journals of the 1830s and 1840s are full of sound, fury, and bewilderment. Although most chemists were certain that organic chemistry was a respectable part of science, they were much divided in their opinions on the nature of the beast.

The development of organic chemistry gave rise to a continuing debate that took place at two distinct levels. The first and fundamental question was, granted that there is such a thing as organic chemistry, loosely defined, what is the proper basis for exploring and describing the field? From which branch of chemistry should analogies be drawn to develop the whole of chemistry? Should organic or inorganic chemistry be regarded as better understood, as simpler, as a more suitable foundation from which to grope for knowledge among the murky thickets of the chemical forest?

The conservative answer was provided by Berzelius. 'In exploring the unknown, our only safe plan is to support ourselves upon the known.' Where chemistry was concerned, the known was the inorganic realm, where the properties of compounds were explained in terms of the properties of their constituent elements. After all, a mineral was what it was precisely because of its elemental constitution—this seemed self-evident to Berzelius. Atoms owed their properties to their electrical natures; thus, when Berzelius said that we should support ourselves upon the known, he meant that electrochemical dualism was to be the model for organic investigations.

There were reasonable grounds for opposing Berzelius, resting on a denial of the truth of electrochemical dualism even in the stronghold of inorganic chemistry. Electrochemical dualists believed that atomic properties, the powers, forces, or natures associated with individual particles of matter, were the essential factors to be considered in the search for an explanation of chemical phenomena. Chemists who rejected these ideas often suggested that arrangement was in principle the fundamental concept that led to conceived connections between chemistry and crystallography. The overall debate may be broadly considered as taking place at two levels, one concerned with the direction in which chemical analogies should be applied, the other with the relative importance of atomic natures and positions. Although both levels of the debate had repercussions for the development of theories of chemical affinity, and were, indeed, frequently inextricably intertwined, the second level, 'power versus arrangement', is more immediately relevant to our theme.

Three men, Dumas, Laurent, and Gerhardt, were particularly influential in countering the sway of electrochemical dualism. Dumas began as a follower of Berzelius's ideas, but came to oppose them, upholding instead a relational theory of chemical types. His influence not only weakened Berzelius's position but also, by drawing attention to the extreme complexity of chemical affinity, discouraged chemists from studying it. Laurent also proposed a relational theory, owing much to crystallography. The effect was once again to distract interest from problems of atomic natures, and to lend support to those who stressed the role of arrangement in chemical phenomena. Chemical affinity, regarded in terms of atomic powers, was the loser; but the structural side of affinity, valency, was to gain from extensions of ideas like Laurent's.

Charles Gerhardt collaborated with Laurent, gaining greatly thereby —intellectually, but not professionally. There is a structural element in his thought, and his interaction with Laurent helps one to understand the latter's development. His main contribution, however, from the standpoint of affinity theory, was his advocacy of a relative, rather than fixed and absolute, electrochemical dualism. The acceptance of his ideas, incorporated with some of Laurent's, facilitated the mutual modification and reconciliation of electrochemical dualism with theories of types. This in turn opened the way to the re-admission of atomic properties into chemical theorizing, while still rendering affinity tables unreliable for predictive purposes. There was by 1860 a clear awareness that the problems of chemical affinity could be subdivided into a question of atomic forces or chemical energetics on the one hand, and a structural aspect, valency or atomicity, on the other.

Chemists in the 1860s and 1870s were much involved with thermochemistry, chemical thermodynamics, structural chemistry, and valence theory—all facets of the anterior blanket concept of chemical affinity. There is much in this chapter on organic chemistry that has little obvious connection with affinity theory; the connection is there throughout,

A M—M

but is frequently historical and not immediate. Developments in organic chemistry in the middle years of the nineteenth century provide the conceptual bridge between the earlier notion of electrochemical affinity as a power of matter, an atomic attribute, and the later concepts of valency and energetics. The bridge, however, is scarcely straight!

Dumas

In 1831 Dumas wrote an open letter to Ampère,[1] in which he stated that electrochemical dualism reigned equally in the organic and inorganic kingdoms, and was responsible for the arrangement of atoms within their molecules. Three years later he asserted, 'All of modern chemistry is based on the idea of duality among substances, which is in admirable agreement with electrical phenomena.'[2] He was, at this date, almost convinced that the electrochemical theory was true, asserting it more dogmatically than Berzelius had ever done. Dumas's attitude to dualism is all the more striking in the light of his methodology,[3] according to which theories were important in proportion to their usefulness, and were regarded 'as something quite distinct from truth'.[4]

In 1836, in his *Leçons sur la philosophie chimique*, he discussed modern theories of affinity. Electrochemical theories were the only ones he seriously considered, and there were three main contenders among them, proposed by Ampère, Davy, and Berzelius respectively. Dumas was initially captivated by Ampère's ideas but in 1836 he rejected them because he thought they failed to account for certain instances of chemical combination; sulphur, for example, could combine with copper, which was electrically more positive, and with oxygen, which was more negative. As we have seen, however, Ampère's theory was very well able to take care of such difficulties. Davy's theory, although satisfactory chemically, relied on contact electrification, which Dumas looked on as discredited. This left him with Berzelius's theory, which alone had thus far withstood the attacks of chemists and physicists alike. However, although Dumas at this point regarded the electrical nature of atoms as the principal factor in determining chemical affinities and hence reactions, he did allow a certain influence to arrangement also. He objected strongly to the prevalent notion of indivisible chemical atoms, and was convinced, with Ampère, that these were in fact integrant molecules. The different arrangements of their component atoms were

[1] *Ann. Chim. Phys.* **47**, 324–35 (1831).

[2] *J. Pharm.* **20**, 261 ff. (1834).

[3] cf. Satish C. Kapoor, 'Dumas and Organic Classification'. *Ambix* **16**, 1–65, esp. pp. 2–6 (1969).

[4] J.-B. Dumas, *Leçons sur la philosophie chimique, recueillies pari M. Bineau*, p. 60. Paris (1837). The *Leçons*, although published in 1837, were given in 1836.

responsible for the phenomena of elemental allotropy and of polymorphism in general. Thus differences in arrangement could lead to differences in chemical properties, presumably including chemical affinities.[5]

In the same year, he and Liebig published a note on the present state of organic chemistry, accepting as a physical fact the implications of their methodological approach.[6] Dumas, like Berzelius, had long viewed the use of inorganic analogies as necessary to the development and elucidation of organic chemistry.[7] While Berzelius stressed the analogies between the branches of chemistry, he still thought that they were, at least for the moment, distinct. Dumas's dualism went further, making of them but one chemistry. In 1834 Dumas had stated that all his efforts sought to transport dualism into organic chemistry.[8] Now, in 1837, he and Liebig were confident that compound radicals in organic chemistry were in fact the analogues of inorganic elements. 'In mineral chemistry, the radicals are simple; in organic chemistry, the radicals are compound, that is all the difference. The laws of combination, the laws of reaction, are otherwise the same in these two branches of chemistry.' Dualism in organic chemistry and the radical theory seemed mutually supporting and well established. When Dumas first propounded his 'theory or law' of substitutions, he did so in support of dualism, and wrote in terms of simple replacement ratios, which could, in the vast majority of cases, be expressed as volume for volume. For example, the action of chlorine on alcohol, with the production of chloral, involved the substitution of a simple volume ratio of hydrogen for chlorine. In Dumas's terms, alcohol was $C^8H^{12}O^2$, chloral was $C^8H^2O^2Ch^6$; alcohol had lost H^{10} and gained Ch^6.[9]

In the initial version of the theory of substitution, only ratios were mentioned, and Dumas made no comment on the effect on the radical into which substitution took place.

. . . I have never said that the new substance made by substitution would have the same radical; the same rational formula as the first. I have said exactly the opposite on hundreds of occasions . . .

The Theory of Substitutions expresses a simple relationship between the hydrogen evolved and the chlorine absorbed. . . .

But if you have me say that the hydrogen removed is always replaced by the electronegative substance, you attribute to me an opinion which my research on indigo disproves, because the hydrogen lost by white indigo is not replaced by oxygen when it is converted to blue indigo, as I showed some time ago . . .

And if you have me say that hydrogen is replaced by chlorine which plays

[5] ibid., p. 308.   [6] C. r. Acad. Sci., Paris 5, 567 (1837).
[7] See Kapoor (op. cit. (3), 18) for Dumas's use of analogy.
[8] J. Pharm. 20, 262–3 (1834).
[9] Mém. Acad. Sci. 5, 519 (1838).

the same part, then you attribute to me an opinion against which I strongly protest . . .[10]

Berzelius had accused Dumas of holding just this opinion, which contradicted both the former's regulative principle, and, specifically, his notion of affinity.[11] That he was right in attributing this idea to Dumas clearly emerged after the latter's work on chloracetic acid,[12] where he extended the theory or law of substitutions to a theory of chemical types; in certain compounds, Dumas claimed, substitution of hydrogen by chlorine did not alter the essential qualities of the type.[13] Clearly these ideas were, as he recognized, difficult if not impossible to reconcile with the electrochemical dualism that he had so dogmatically affirmed only one year previously. 'But,' he asked,

are these electrochemical ideas, this special polarity attributed to the molecules of simple substances, based on such evident facts that they must be blown up into articles of faith? At least, we must recognize them as hypotheses; have they the properties of explaining and predicting the facts with such reliability that chemical research relies on them for its help?
One must admit that they have not.[14]

Dumas had thus progressed from confident support of electrochemical dualism to a position where he doubted even its value as a methodological tool, and ignored it in propounding his new theory. He had not simply changed his mind overnight, but had observed his own dictum that '[t]heories . . . oblige us scrupulously to study all the facts which contradict them, until it is proved that the facts have been inadequately observed, or that the theory must be modified'.[15] Dumas had long found dualism plausible and fruitful for organizing chemical information; he adhered to it while it fulfilled this role for him.[16] Gradually, however, he became aware of various anomalies of dualism; by 1839 he thought he had found a better theory, and so rejected dualism.[17]

Berzelius identified this change as nothing less than a complete revolution in the state of chemistry. He pointed out that Dumas advocated two great principles in chemistry, one of which, isomorphism, was valid in the inorganic realm, while the other, substitution, was valid for organic chemistry. Berzelius believed that Dumas's first sin was his acceptance of substitution, a principle irreconcilable with electrochemical dualism. Dumas's second sin, according to Berzelius, was a belief that substitution and isomorphism 'derive from the same cause, and that they will be able, with time, to be generalized under a common

---

[10] *C. r. Acad. Sci., Paris* 6, 647 (1838).   [11] ibid., 629 (1838).
[12] ibid., 8, 609 (1939).   [13] ibid.   [14] ibid.
[15] *J. Pharm.* 20, 288 (1834).
[16] e.g. *Traité de chimie appliquée aux arts*, vol. 1, p. liii. Paris (1828).
[17] See Kapoor, op. cit. (3), 13–24.

expression'.[18] Dumas was hurtling towards a unified chemistry more rapidly than Berzelius's caution could approve.[19] He did indeed already consider substitution and isomorphism as related processes or principles.[20]

Berzelius, having realized that his approach was incompatible with Dumas's, at once set about attacking the latter and restoring the *ancien régime*. He had first to dispose of Dumas's ideas about chloracetic acid. Berzelius observed that acetic acid was much more like formic acid than like chloracetic acid—so much for the virtue of types! Basing his argument on the radical theory, which had recently received tremendous support from the isolation of cacodyl, Berzelius suggested that chloracetic acid, so called, was in fact nothing more than a combination of oxalic acid with a chloro-carbon corresponding to oxalic acid. This approach had the virtue of returning the new acid into a class of bodies that was already known. Berzelius concluded:

> We have reached a point where we are beginning to see a theory of organic compounds; but if, instead of letting this develop as our experience grows, we want to base it on isolated facts, considered without regard for their relations with the general system of our knowledge, and by giving explanations which do not harmonise with the principles of the science, and if, moreover, we want to conclude that this lack of agreement must lead us to reject as erroneous principles which are already well established on other grounds, then we shall never succeed in finding the truth.[21]

Berzelius was not the only one to understand the revolutionary significance of Dumas's approach. Auguste de la Rive was another who realized that the acceptance of the new theories would mean the complete subversion of the old ideas about affinity. This distressed him, so that he wrote to Dumas, 'I cannot help thinking that there is something well-founded in the table of the relative chemical powers of bodies.'[22] He could not accept the complete denial of all previous ideas about chemical affinity as a principal determinant of chemical properties. Nor, it would appear, could Dumas himself, for, although he now asserted that the situation of particles was primarily responsible for their properties, he nevertheless continued to lecture on affinity as a specific molecular force, even though he refused to commit himself to any firm ideas about its nature.[23] In addition, he agreed that the electrical nature of bodies still had some influence:

only it must be agreed, that it is at the moment when the combinations are

---

[18] *Ann. Chim. Phys.* **71**, 137 ff. (1839).   [19] cf. p. 161 above.

[20] *Phil. Mag.* (3) **17**, 185 (1840); *Ann. Chim. Phys.* **73**, 99 (1840).

[21] op. cit. (18).   [22] *Phil. Mag.* (3) **17**, 183 (1840): letter of 25 Oct. 1839.

[23] *Académie des Sciences*, Paris MS. Dumas, 10: Chemistry course of 1839–40 at the Sorbonne.

made, at the moment when they are destroyed, that the function (rôle) of electricity may be observed.

But when the elementary molecules have taken their equilibrium, we no longer know how to define the influence that their electric properties may exercise, and no one has set forth views on this subject which agree with experience.[24]

Dumas was thus advocating a theory that was primarily unitary and relational. His classification and theory were closely connected, and based upon the relations between atoms in integral molecules. He still, however, admitted that electrical natures also influenced properties. Just how the electrical natures of bodies operated was, however, obscure. Prediction and even understanding of chemical phenomena were therefore not helped by the vestiges of dualism, which were sterile and thus irrelevant to the progress of chemistry. Chemical types instead would provide the basis for such progress.

In a paper of 1840 on chemical types, Dumas gratefully cited Liebig's approval of his theory of substitution,[25] referring to the substitution of chlorine for manganese in permanganic acid. The principles arrived at from the study of organic phenomena were now being used to account for inorganic reactions. The direction of the analogy had been totally reversed.

Liebig's approval was, however, qualified. He still retained the notion that the order of reaction was determined by electrochemical affinities; isomorphous replacement and substitution 'depend[s] entirely upon the chemical force which we call affinity'. Thus this apparent support for Dumas's theory was also a reassertion of Berzelius's theory of affinity. Liebig was here applying the organic–inorganic analogy in both directions simultaneously. This meant, of course, that he was open to attack from both sides of the fence; his comments on Dumas's theory were followed by the sarcasm of S. C. H. Windler's paper on the spun chlorine fabrics on sale in London shops.[26] This paper was directed equally at Dumas's assertion that

In an organic compound, all the elements can . . . be successively displaced and replaced by others. Those which disappear the most readily, after abstracting certain conditions of stability which are still unknown, are those with the most energetic affinities.[27]

Even when going so far as to allow carbon substitution within a unitary system, effectively removing the barrier between organic and inorganic

---

[24] *Phil. Mag.* (3) **17,** 188 (1840).   [25] *Phil. Mag.* (3) **16,** 446 (1840).
[26] *J. Chem. Pharm.* **33,** part iii: translated *Phil. Mag.* (3) **17,** 75, (1840). The paper was written by Wöhler, and after it, Dumas dropped the idea of carbon substitution that is implied in note (27).
[27] *C. r. Acad. Sci., Paris* **10,** 156 (1840).

compounds, Dumas was unable to disembarrass himself of the notion of chemical affinity. His ideas on arrangement, however, made it impossible to obtain any precise ideas about the role and nature of chemical force. There might even be some truth, though without present heuristic value, in dualism: 'I am far from denying that chemical and electrical forces may be the same.'[28]

Dumas urged, perhaps as a merely formal sop to his critics, that a synthesis of aspects of electrochemical dualism with the theory of types would provide the key to chemical phenomena.

Pushed to extremes, each of them [i.e. dualism and types] according to my opinion, would find itself leading to absurdity. Ruled by experiment, and restrained by it within sensible limits, each of them will have a large part to play in the explanation of chemical phenomena; and . . . I believe that in chemistry the nature of molecules, their weight, their form and their situation, must each exercise . . . a real influence on the properties of bodies.[29]

In fact, however, Dumas limited his arguments to the role played by arrangement; this limitation may have been propagandist, but he probably believed, as he frequently stated, that 'it is above all from their situation that particles derive their properties'.[30] This belief referred to a theory that, even if 'distinct from truth', had at least heuristic potential. No more could be expected, for the problems of chemical reaction were too complex for full treatment.

Since situation was so important in his chemical theory, it will be useful to glance at his ideas on this topic. Some of the force of the objections raised above against his bid to retain the vestiges of electrochemical dualism within a theory of types is lessened, as the reasonableness of his confession of ignorance of the role played by electrical characters within a stable molecule is strengthened, when one looks at his picture of molecular constitution.

Dumas inherited the atomic theory, which he at first accepted. It was useful for relating chemical phenomena in a systematic way, and had also served predictively. Subsequently, his failure to distinguish between atoms and molecules led him to believe that atoms were not indivisible; he therefore rejected Daltonian atomism, proposing instead two kinds of particles, physical and chemical atoms. The former were physically indivisible, though chemically divisible, and they obeyed Ampère's hypothesis. Chemical atoms were the smallest parts of elements that were involved and conserved in chemical reactions. Dumas thought that the existence of the former could be empirically verified, while the existence of the latter could not. Chemical atoms thus became metaphysical entities, and so were methodologically unsatisfactory.[31]

[28] ibid., **10**, 177, 178.   [29] ibid.   [30] ibid., **10**, 171.   [31] See Kapoor, op. cit. (3), 7–13.

Dumas looked for alternatives to the 'metaphysical' atomic theory. His theory of types dealt with aggregates, rather than with individual particles. So too did the works of C. L. Berthollet, to which Dumas owed a general intellectual debt;[32] and Berthollet had suggested that there was a broad analogy between chemical and gravitational attraction.[33] Such ideas may have contributed to the planetary theory of matter which Dumas again put forward in 1840, having first suggested it in 1836.[34] He supposed that atoms did indeed have electrical properties. But atoms were not arranged in static configurations, analogous to crystal structures; they were instead components of dynamical systems in equilibrium. Dynamical here does not refer to the force theories of matter discussed in earlier chapters, but rather to something analogous to the pattern of planets in our solar system. The particles were held in dynamical equilibrium by 'the diverse molecular forces whose resultant constitutes affinity', and even if electrical forces were part of this scheme, it was clear that there would no longer be room for the static dualistic schemes of Lavoisier and Berzelius. When one particle within such a system was replaced by another particle of a different kind, a new equilibrium would necessarily be established, and the closeness of this equilibrium to the original would determine the closeness of the reactions of the product to those of the reactant molecule.

The picture is vague and qualitative, and Dumas clearly had no intention of using it either for prediction or for detailed explanation; perhaps its greatest utility was in its demonstration of how complicated the system of molecular arrangement and molecular forces might be, and in its removing from immediate consideration problems arising from a consideration of the system. Its more practical implication was that, when minimal changes in properties indicated that the type had been conserved, that is that the molecular and atomic arrangement had not altered in the course of substitution, then it was legitimate to infer that the new compound possessed an analogous composition to the old one. In general, chemical types embraced bodies that contained the same number of equivalents united in the same manner, and that possessed the same fundamental chemical properties. In deciding the probable course of chemical reactions between substances of the same chemical type, one could reasonably use analogical argument, coupled with a knowledge of the general forces of chemistry.[35] Since, as we have seen, Dumas was always prudently vague about chemical forces, the latter recommendation was not enormously helpful.

---

[32] Dumas, op. cit. (4), pp. 379–80.   [33] See Chapter 7.

[34] *Ann. Chim. Phys.* **73**, 73 ff. (1840). Dumas put forward an early version of the planetary hypothesis in 1836 (op. cit. (4), p. 234).

[35] *C. r. Acad. Sci., Paris* **10**, 159 (1840).

Dumas's planetary theory of matter has been used to illustrate his widening divergence from the bases of electrochemical dualism: preformation, electrically polar chemical atoms, the binary rather than unitary composition of molecules, and a system whose properties were determined by individual atomic natures, rather than by relations within a unified structure. The motives that had led Dumas to propose the planetary theory persisted, though he soon abandoned the theory itself.[36]

In 1868, when the early dust had settled, Dumas published a retrospective account of theories of chemical affinity, and commented upon them.[37] His account was fairly long, unusually so for the *Académie*'s *Compte Rendu*, but the editors noted that they were publishing it in full because of its great importance.

Throughout his career, Dumas had speculated on the possible connection between chemical affinity and Newtonian attraction[38]—just as Faraday and Davy had rightly seen their electrical ideas as a continuation of certain aspects of the Newtonian tradition. Dumas recognized that Berthollet's work had been Newtonian in its inspiration, and considered that his chemical statics had led to a partial elucidation of the laws of molecular movement. Then came Davy's electrochemical theory, and his isolation of the alkali metals. Thereafter, electrical theories had borne no fruit for chemists, until Ampère proposed his theory, which was indeed electrical, but could be interpreted more widely as uniting crystallography with Newtonian attraction. It is worth noting that to have understood this, and to have approved of it, Dumas must have had some sympathetic understanding of the dynamical character of Ampère's theory of matter[39]—indeed, Dumas had once before espoused a form of point atomism.[40]

Dumas observed in 1868 that Ampère's work had excited little interest at the time of its appearance. It represented neither the absolute ideas of affinity as this concept was then understood, nor the dualism of the time, which gave Lavoisier's nomenclature all the force of a doctrine, and saw in language 'a real representation of the intimate constitution of bodies ... Today we are disposed to admit that the theory of chemical combination proposed by Ampère best satisfies the general laws of mechanics'; this was because it was based on universal attraction and the special laws of chemistry, and took account both of the shape and the complexity of molecules. Since Ampère's system, modified

[36] The theory was published in *Ann. Chim. Phys.* for Jan. 1840, and abandoned by 3 Feb. 1840, when he read his second paper on types to the *Académie* (*C. r. Acad. Sci., Paris* **10**, 149–75 (1840)). I am grateful to Dr. N. Fisher for pointing this out.

[37] *Ann. Chim. Phys.* (4) **15**, 70 (1868).

[38] For example, see note (23) above.

[39] See note (43) below.   [40] Dumas, op. cit. (4), p. 230.

by Gaudin's ideas on the use of symmetry considerations in deciding on formulae,[41] was concordant with French nomenclature, Dumas thought it not surprising that it had been finally accepted by many chemists who were obliged to consider the phenomena of substitution. As a result, Ampère's theory 'has at the same time rendered the search for a precise electrochemical theory less ardent. . . .

One is also led to conclude that at the minute, . . . chemists are scarcely concerned with the search for an electrical theory of affinity.' Instead, said Dumas, the theory of types was being used more and more,

and all hypotheses on the interior arrangement of the elements (within molecular types) have been abstracted . . .

At the same time one is brought back to the thought which attributes to the molecules of compound bodies a more complex constitution than could be derived from binary nomenclature, and which looks upon them as planetary or crystallographic systems, offering the synthesis (*réunion*) of several atoms or centres of force, moving in the one case, fixed in the other.

Thus one finally returns to the thought which connects affinity directly to universal gravitation.[42]

This did not mean for Dumas that dualism, in the sense of molecular antagonism leading to combination, represented anything other than the true state of affairs; but once bodies had combined, French nomenclature said nothing about whether the two bodies kept their distinct characters, or were fused in a complex system. This is where Berzelius went beyond Lavoisier, and where he overstepped the mark. The *Académie* was right and had nothing to regret in its adherence to Lavoisier's language and to Newtonianism.

---

[41] Gaudin's theory (*Ann. Chim Phys.* **52,** 113 (1833); *C. r. Acad. Sci., Paris* **45,** 920; and there are related papers in *C. r. Acad. Sci., Paris* **25,** 664 (1847); **32,** 619, 755; **34,** 168) deals with the physical arrangement of atoms. Basically, he believed that he had discovered the true primitive form of bodies, together with the law presiding over the arrangement of their atoms; molecular form accorded with the solids obtained by cleavage. Real atoms (*C. r. Acad. Sci., Paris* **45,** 920 ff.) combined in geometrical arrangements, symmetrically disposed by their relative position which holds them in equilibrium under the action of universal weight [*pesanteur*], As a result, the reason for combination is a *geometrical reason,* just as much as the aggregation of molecules in crystals is, with the following difference: simple atoms, spheroids doubtless engendered in a cubic system, have, in themselves, no determinate form to contribute to that of the molecule.

All molecules, without exception, are formed from linear molecules arranged parallel to one another

and by the parallel grouping of linear molecules along their three axes, one could build all the geometrical polyhedra of regular bodies.

A more detailed account of Gaudin's atomic structural theory is given by S. H. Mauskopf, *Isis* **60,** 61–74 (1969).

[42] *Ann. Chim. Phys.* (4) **15,** 89.

Thus Dumas concluded his account of affinity. In some ways, it is an exaggerated summary of the developments that will be considered in this and the next chapter. The most striking thing about it, from the standpoint of this chapter, is its explicit statement that the theory of types as seen by Dumas led to a recognition of the complexity of the problem of chemical affinity, and to a corresponding lessening of enthusiasm for, and confidence in, the search for a detailed theory of affinity within the chemists' fold. The problem was not argued away, however, but subsumed under a wide-ranging mechanical scheme, and relegated to the realm of molecular physics, where most chemists left it, to be supported by thermochemistry and to await the advent of thermodynamics. On the other hand, the recognition that structural factors were relevant to the study of chemical affinity was largely responsible for the rise of valence theory and stereochemistry. Affinity became fragmented as a topic, and removed from the mainstream of laboratory chemistry.[43]

With this reminder of the role of structural factors, Auguste Laurent may be introduced.

Auguste Laurent

In 1836 Laurent entered the debate with his doctoral thesis,[44] presenting a new theory of organic combination while criticizing the two main theories then current: electrochemical dualism, involving preformation, and the unitary theory. The structure of Laurent's thesis is tripartite. First come objections to current theories. These objections are in turn overcome by criticism. Finally, Laurent produces his own theory, salvaging aspects of dualism (though not electrochemical dualism) and of the unitary theory, and welding these parts into his own creation.

To the theory of electrochemical dualism, Laurent raised two distinct objections of his own, and another put forward by his contemporary, Baudrimont. Baudrimont's objection was that chemical reaction involved the movement of atoms, altering their relative positions. How, then, could one infer anything about the relative positions of atoms in a compound from the reactions which the compound underwent?

Laurent's first objection was that the ideal chemical formula should express the totality of reactions undergone by a compound. But most chemical substances behave differently with different reagents and

[43] The substantiation of this paragraph will be the burden of the remainder of this book.

[44] *Académie des Sciences*, Paris MS. Laurent, 1836. Part of this was published in *Ann. Chim. Phys.* **61**, 125 (1836); the remainder is now available in *Bull. Soc. chim. Fr. Docum.* 31 (1954), edited and introduced by J. Jacques. Dumas was one of Laurent's examiners.

under different conditions, so that it was hard to select a satisfactory formula from the available alternatives. Although Laurent did not know it, the same difficulty troubled Berzelius.

Laurent's second objection, distinctive of his approach to chemistry, was based on crystallographic arguments. Many French crystallographers at this date still accepted Haüy's ideas, or others derived therefrom, about crystal structure, postulating macroscopic crystals built up from molecular groups of related regular geometric form.[45] The basic building blocks were few in number—Haüy proposed three 'primitive forms'.[46] For example, suitable arrangements of cubic nuclei could yield dodecahedra, pyramids, or other forms. If one accepted this picture of crystal structure, then dualism, with its implicit concept of preformation, raised a serious difficulty. How could two separate molecules, with different crystalline forms, combine to give a third molecule whose crystalline form might well be incompatible, in terms of symmetry, with its constituents? The problem, earlier touched upon by Ampère, was one troubling the mineralogists in their search for a classification of mineral species, but generally ignored by chemists.[47] James Dwight Dana gave a clear statement of the problem in his *System of Mineralogy*, pointing out that compound molecules similar in form to their crystals

cannot be formed by the *juxtaposition* of atoms. An atom of sulphur, the primary of which is the *rhombic octahedron*, united to an atom of lead, whose primary is the *regular octahedron*, could not in any way be made to receive the cubic form of galena; nor, were the molecules equal spheres, would it be a less difficult task—at least eight equal spheres would be required.[48]

In other words, Haüy's kind of crystallography could only work with preformation in chemistry when modified to admit compound units. Laurent came to adopt this conclusion, employing compound elemental 'atoms'.[49] Since he had worked under Dumas, who was interested in Ampère's ideas, a possible chain of influence is strongly suggested. This suggestion is strengthened by Laurent's reliance, after the mid 1840s, upon the hypothesis of Ampère–Avogadro; he used this hypothesis in working out chemical formulae as a tool for classification.[50] In 1836,

[45] J. C. Burke, *Origins of the Science of Crystals*, p. 86. Berkeley and Los Angeles (1966).

[46] R. J. Haüy, *Traité de Minéralogie*, p. 31. Paris (1801). This is discussed by D. C. Goodman, 'Problems in Crystallography in the Early Nineteenth Century'. *Ambix* **16**, 157 (1969).

[47] There were exceptions to this generalization, both prior to and after Laurent, e.g. F. S. Beudant (*Q. J. Sci. Arts* **6**, 117 (1819)) was concerned with the connection between crystalline form and chemical compositions, while Dumas (*Ann. Chim. Phys.* (4) **15**, 93 (1868)), was still worried about the relation of crystalline form to atomic arrangement.   [48] J. D. Dana, *System of Mineralogy*, p. 76. 2nd edn (1844).

[49] See note (61) below.

[50] e.g. A. Laurent, *Chemical Method*, trans. W. Odling, pp. 65–6, 68. London (1855).

however, Laurent seems not to have advocated compound atoms as a solution of the difficulty. Nor was he willing to accept the notion that in crystals the atoms were symmetrically disposed, without regard to their possible modes of combinations—this would be sterile for chemistry, however much it simplified the crystallographer's problems. Strict electrochemical dualism was untenable, for the reasons already given; but there was some evidence of the existence of regularly occurring groups within molecules—for example, Laurent thought that the constancy of colour through a series of salts of the same metal was due to the persistence of a particular group of atoms.

So far, Laurent had merely criticized existing arguments; having cleared the ground, he was ready to propose his own theory, a compromise between electrochemical dualism and the unitary theory. He stated his thesis compactly:

When a molecule, containing atoms grouped according to a certain form, is confronted with another molecule, the atoms influenced by the presence of the new molecule adopt another arrangement, but without being withdrawn from their reciprocal influence; the same is true of the atoms of the new molecule.

It followed that 'in combining, the two molecules must thus lose their form, *but* the atoms of one of them do not pass into the other'. Laurent's detailed exposition of his thesis explains this passage. The model he offers for describing organic combination is a structural one, related to crystallography. He regards organic molecules as made up of a central radical, the structural kernel, surrounded by an external complex. The kernel or nucleus of the molecule determines the group of compounds to which the molecule belongs; if substitution occurs within the nucleus, then the class to which the compound belongs is changed. The external complex, the superstructure of the molecule, however, may undergo substitution—for example, a hydrogen atom outside the nucleus may be replaced by a chlorine atom—without any concomitant change in chemical class or type.[51] This model gave a useful basis for a chemical classification. It also implied that the place of the individual atoms within a molecular structure was more important than the nature of these atoms.

Laurent's structural model, unlike Dumas's planetary one, was dualistic, clearly distinguishing the nucleus from the less stable superstructure. His types underlay his classification, and in so far as these types conformed to a structural dualism, his classification also was dualistic; and classification, organizing knowledge and indicating

[51] This is all taken from Laurent's thesis; see note (44) above.

relations, was the crux and goal of chemical endeavour for Laurent, and the major problem in organic chemistry at that time.

Electrochemical dualists had found a basis for classification in the affinities of elements. Individual atomic natures determined the properties of radicals and of molecules, and were used by Berzelius in framing his classification of minerals. Laurent's dualism, however, was a structural but not substantialist dualism, largely ignoring those attributes that distinguished the atoms of different elements.

There were, then, both similarities and major methodological differences between Laurent's and Berzelius's approaches to organic chemistry. But dualism is not the only, or even the principal, point at issue. Laurent and Berzelius both assumed, as a working rule, that there was some analogy between organic and inorganic chemistry; they differed with respect to the direction in which this analogy was to be applied, Berzelius starting out from the inorganic realm, Laurent from the organic. In both cases, the analogies were pursued as likely to be fruitful for chemistry. Thus far, the division is clean and clear.

The picture becomes somewhat complicated by Laurent's indebtedness to the French crystallographic tradition for his ideas on organic combination; this tradition was an inorganic one, though not in the line of Berzelius's dualism. Furthermore, Laurent himself constantly referred to inorganic substitution or isomorphous replacement as closely comparable with organic substitution.[52] Since the former had been known for more than a decade before Laurent developed his ideas, it is reasonable to suppose that he drew his analogy from the prior theory; in which case, he *was* using analogies from inorganic chemistry for guidance in organic chemistry. Berzelius was unhappy when he came to read Laurent's papers, because of the latter's neglect of the special analogies from electrochemical dualism.

Laurent promptly used his theory of substitution or of derived radicals, based as it was on a geometrical model, to account for a wealth of experimental facts, many of which he had discovered himself. For example, he used the theory to explain why, in the chlorinated compounds of naphthalene, one could isolate compounds containing as many equivalents as the hydrocarbons whence they derived, while one could not isolate the hypothetical radical required by Liebig and other electrochemical dualists.[53] Laurent was to insist repeatedly that organic substitution could be described in terms of the *isomorphism* of different substituents, such as that of chlorine with hydrogen, hydrogen with ammonium.[54] This emphasis led to his use, ever more confident, of

[52] e.g. *C. r. Acad. Sci., Paris* **20**, 357–66 (1845).
[53] *Ann. Chim. Phys.* **66**, 314–35 (1837).
[54] e.g. *C. r. Acad. Sci., Paris* **11**, 876 (1840); **12**, 1193 (1841).

structural models. In 1840, in one of his innumerable reclamations against Dumas, he pointed out that his theory differed greatly from that of Dumas. In the latter, 'there is no question of the place which chlorine or oxygen should occupy in the new combination . . .'[55]

Laurent believed that structure determined reactivity, and was directly connected with crystalline form. Papers continued to flow from him on this connection and on the connection of form with constitution.[56] Systematic arrangement, recognizing these connections, produced a simple classification of all substances, precisely similar to that of mineral chemistry—given Laurent's own classification of mineral substances![57] Laurent and Berzelius started to correspond in 1843, both giving full and revealing statements of their positions. Laurent began by seeking to persuade Berzelius that they really shared a great deal of common ground; both of them, for instance, were definite in their refusal to accept the idea of substitution in relation to Dumas's mechanical types, substitution products with similar chemical formulae but differing in chemical properties.[58] If these types existed, then further substitution into them should yield predictable products, which were not in fact produced. Laurent was even ready to agree with Berzelius that substitution did lead to some change in properties, even when the type was conserved. In small molecules, an isomorphically introduced substituent might produce a large mechanical imbalance, with a corresponding change in properties. Large molecules could absorb the mechanical disturbance much more readily, and hence isatin and chlorisatin, for example, scarcely differed in their chemical behaviour. Berzelius, in return, admitted to some common ground, but insisted that he could not consider 'chlorine as having adopted the nature of hydrogen, since chlorine always will remain chlorine', even though this retention of individual atomic electrical character might be masked by the phenomena that were classed under the heading of copulation or conjugation.[59]

[55] *Académie des Sciences*, Paris MS.: Laurent to the President of the *Académie*, Bordeaux, 12 Feb. 1840: 'Il n'y est nullement question ni de la place que doit occuper le chlore ou l'oxigène dans la nouvelle combinaison . . .'

[56] e.g. *C. r. Acad. Sci., Paris* **15**, 350–2 (1842).

[57] *C. r. Acad. Sci., Paris* **17**, 311–12 (1843).

[58] Söderbaum, ed. *Jac. Berzelius Bref*, vol. 7, pp. 181–5 (6 vols. + 3 suppl., Uppsala, 1912–32): letter of 12 May, 1843.

[59] ibid., vol. 7, pp. 188 ff.: letter of 20 Oct. 1843. Copulated or conjugated compounds were binary compounds or terms 'in which one of the terms, the active compound, conserves its property of uniting with other bodies, whilst the other term, which we call the copula, has lost all tendency to combination, with certain exceptions'. (Berzelius, *Traité de chimie*, vol. 5, pp. 26 ff. (Paris, 1949), cited in translation by J. R. Partington, *A History of Chemistry*, vol. 4, p. 374. London (1961–4).) For example, in chloracetic acid the active term was supposed by Berzelius to be oxalic acid, $H^2O.C^2O^3$, responsible for the compound's acidic properties. It was copulated with the passive term, C^2Cl^6, wherein chlorine was relatively inactive, because firmly bound to carbon.

It was probably then that Laurent read Berzelius's memoir on the allotropy of elemental substances, which referred to different states of the elements.[60] Laurent considered that these different 'atomic' states, responsible for different chemical reactivity—for different degrees of affinity, as Berzelius would have phrased it—were due to the molecular nature of elemental 'atoms', so called; different arrangements of the constituent particles would result in different properties.[61] Berzelius, in contrast, based his methodology on the notion of atomic properties derived from experiments in inorganic chemistry. He therefore refused to believe that affinity could arise from structural factors derived from the obscure depths of speculations about organic compounds. He did not deny the potential usefulness of Laurent's approach, but insisted that, since it was diametrically opposed to his own, there would be no point in pursuing the correspondence:

The views which you have just communicated to me in this letter have persuaded me that it would be useless to try to reconcile our theoretical opinions with one another, since we have completely opposite points of departure. As far as I can judge from your memoir on the naphthalic series . . . you are trying to reform the theory of inorganic chemistry according to the ideas you have obtained from your experiments on organic compounds. In my chemical endeavours, I have preferred the diametrically opposed route, basing speculations with regard to organic combinations on theoretical ideas more or less established for inorganic compounds, and although I consider the latter method as preferable, I am by no means blind to the extension of theoretical knowledge which may follow from the method which you have chosen. It will therefore be best if we each follow our own route amicably, in the hope that science will draw profit from both.[62]

This letter accurately identified the main methodological difference between Berzelius and Laurent. Laurent's paper on silicates, published in 1849, explicitly confirmed this interpretation.[63] He first contrasted the simplicity of organic nature with the complexity of the inorganic realm, then went on to classify the inorganic silicates according to the principles he had developed in organic chemistry. This procedure had the great advantage of reconciling chemical composition with crystalline form.

Crystallography had greatly influenced Laurent's chemical thought in 1836, and remained significant for him until his death in 1853. His crystallographic and structural models were, however, merely heuristic, and not intended to indicate the true positions of material atoms. Laurent revealed and partly developed his attitudes towards structural

[60] *Jahresber. Fortschr. Chem.* **20**, II, 13 (1841 for 1840): trans. in *Scient. Mem.* **4**, 240–52 (1846).

[61] Söderbaum, op. cit. (58), vol. 7, p. 200: letter of 5 Jan. 1844.

[62] ibid., vol. 7, p. 208: letter of 25 June 1844.   [63] *Institut de France* MS. 2379.

models while interacting with his younger contemporary Gerhardt, a former pupil of the irascible Liebig.

Charles Gerhardt and Auguste Laurent

Charles Gerhardt came to Paris in 1838, bearing a testimonial from Liebig.[64] He enrolled in Dumas's course, while continuing independent research. Liebig wrote frequent letters advising Gerhardt to be prudent in his theoretical opinions, 'for the *Académie* . . . is the implacable adversary of theories'.[65] Even Dumas had at first been hindered by his theoretical tendencies, and Laurent had ruined his future and irritated everyone by the same error. 'For the love of God, don't write about theories, except for German journals!'[66]

Gerhardt was young and innocent, and in 1842, in spite of such warnings, read a paper to the *Académie* on chemical classification, urging that chemical equivalents be modified;[67] the members of the *Académie* were far from pleased. Gerhardt, taken aback at the violence of their reaction, wrote to his friends for advice and sympathy. Liebig replied that he had not read the paper, 'but I've been told that the good impression produced . . . by the importance of the facts was overcome by a lack of caution in the conclusions. Do me the favour of following my advice, which is to refrain from risky theoretical speculations.'[68]

Gerhardt, in his 1842 paper, had aroused hostility by his theorizing. Important in the paper was his belief that organic substances were related in a 'chain of being'. His perception of the key role of series in organic classification derived from this metaphysical conviction, which, in spite of Liebig's warning, he restated in his next major publication, the *Précis de chimie organique*.[69] Gerhardt knew that series were his great insight. 'I think that [homologous series] will be much more useful than substitutions, seeing that it's a matter not only of chlorine and bromine, but of all possible compounds.'[70]

The opening pages of the first volume of the *Précis* presented nature as a system in which chemical affinity and vital force were in dynamical equilibrium. Substitution was accepted as an experimental fact elevated to the general status of a law, while Liebig's theory of compound radicals was dismissed as a secondary offshoot of electrochemical dualism.[71]

[64] Biographical information in this section is taken from E. Grimaux and C. Gerhardt jr., *Charles Gerhardt, sa vie, son œuvre, sa correspondance 1816–1856*. Paris (1900).
[65] ibid., p. 38: Liebig to Gerhardt, 15 Aug. 1839.
[66] ibid., pp. 42–3: Liebig to Gerhardt, 1 March 1840.
[67] *Rev. Sci. ind.* **10**, 145 (1842).
[68] Grimaux and Gerhardt, op. cit. (64), p. 67: Liebig to Gerhardt, 2 Nov. 1842.
[69] C. Gerhardt, *Précis de chimie organique*. 2 vols., Paris (1844–5).
[70] Grimaux and Gerhardt, op. cit. (64), p. 83: Gerhardt to Liebig, 27 July 1844.
[71] Gerhardt, op. cit. (69), vol. 1, p. 10 (cf. vol. 1, pp. 2, 3, 8).

A M—N

One unsatisfactory aspect of the latter was its unwarranted assumption
of preformation, implying some knowledge of the arrangement of atoms
within molecules. Gerhardt, however, argued that nothing was known
about molecular arrangement, and that graphical representation served
merely to indicate analogies in reaction.[72] As he wrote to Wurtz, 'I want
to arrive at laws by seeking to establish equations, general formulae, I
reject all hypotheses about the manner of grouping, etc.'[73] It is important to remember this warning, as otherwise Gerhardt's thought will
appear to be based firmly on structural concepts.

And yet it is impossible not to identify a structural element in his
ideas. For example, when potassium was substituted for hydrogen in
oxalic acid to yield potassium oxalate, there was not much change in

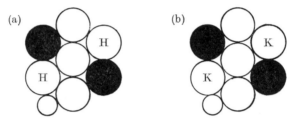

FIG. 6.1. (a) Oxalic acid; (b) potassium oxalate.[74]

properties. One body could replace another *without alteration in the
molecular edifice*.

Gerhardt went on to classify chemical combinations,[75] which fell into
three main categories:

(i) Saline combination, which occurred, for example, when hydrogen
replaced silver in silver acetate. Acetic acid was formed, with
conservation of saline type.

(ii) Substitution in organic chemistry, such as the replacement of
hydrogen in naphthalene by chlorine.

(iii) Substitution by residues, for example,

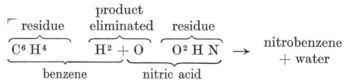

Gerhardt stated that these changes were all effected by affinity between
the reactant elements, but gave no details.[76] Perhaps one reason for his

[72] Gerhardt, op. cit. (69), vol. 1, p. 11. Gerhardt here makes explicit his debt to
Baudrimont.

[73] Grimaux and Gerhardt, op. cit. (64), p. 77: Gerhardt to Wurtz, 9 Jan. 1844.

[74] Gerhardt, op. cit. (69), vol. 1, p. 13.

[75] ibid., vol 1, pp. 59 ff.   [76] ibid., vol. 1, p. 199.

reticence was despair of understanding the precise nature of affinity; chemistry was the science of change, and to be able to describe chemical change one needed to know the composition of bodies. The full concept of composition included the idea of arrangement; direct observation would yield absolute information about the quantitative proportions of bodies, 'but the disposition of the atoms, the way they are grouped, can only be grasped in a relative way'.[77] Analogies, such as those from crystalline structure, were useful, but only in general terms, and could not give any absolute structural information.[78] At most, analogical arguments might suggest that the atoms in one molecule possessed the same arrangement as atoms in another molecule; one could not expect to discover more than this relation between the two structures.

Gerhardt believed in the existence of atoms, even admitting, in the second volume of his *Précis*, that the atoms of different elements did have, relative to one another, differing electrical properties. Sometimes combination took place in a manner that could be correlated with the distance of two bodies from one another on the electrical scale; the opposite, however, was also sometimes true, and this, as Gerhardt noted, directly contradicted the ideas of dualists as they were generally understood.[79] A classification of bodies based on their electrical properties simply could not be absolute, and was therefore valueless for prediction. In other words, Gerhardt objected to the sterility rather than to the wrongness of electrochemical dualism.[80] The same objection was valid against theories of chemical affinity; there clearly was such a thing, but knowledge of the relevant phenomena was, and in the present state of chemistry could only be, insufficient for predictively useful understanding of affinity.

Gerhardt's handling of structural models and concepts changed after 1844, largely in reaction to criticism from Laurent, with whom he now corresponded vigorously.[81] Their fruitful collaboration was one of mutual encouragement and criticism, invaluable to both.

Laurent pounced on Gerhardt's approach to structure as soon as the first volume of the *Précis* appeared.

M. Gerhardt can always reply that not all chemists are agreed on [the way atoms are grouped together in molecules], and that while we are waiting for this [agreement] to come about, if that will ever be possible, it is better for us provisionally to follow a system which is independent of all hypothesis.

[77] ibid., vol. 2, p. 489.
[78] See note (104) below.
[79] Gerhardt, op. cit. (69), vol. 2, pp. 495–6.
[80] C. Gerhardt, *Introduction à l'étude de la chimie par le système unitaire*, pp. 49 ff. (1848).
[81] Their correspondence is published in M. Tiffeneau, ed. *Correspondance de Charles Gerhardt*, vol. 1. 2 vols., Paris (1918).

Nevertheless, M. Gerhardt has not dared to follow his system into all its consequences; indeed, we see that he separates water of crystallization, that he admits that there is a group Az^2O^4, an ammoniacal group, [and] in some cases the hydrochloric acid group and the sulphuric acid group in copulae.[82]

Laurent's heuristic view of structural theories, and indeed of all his theories or models, did not prevent him from believing that compound bodies were really constituted of different groups of atoms.[83] He reified the general concept of structure, but was careful never to go beyond the evidence and extend the process to particular molecular structures. He believed that Gerhardt was too frightened of this excess, and wrote to him in July 1844, 'Say that the arrangements of Berzelius, of Davy, etc., don't satisfy you, but do not reject all attempts directed at the determination of arrangement, since in spite of yourself it dominates you in a multitude of cases.'[84] Gerhardt argued, but Laurent's acute comments gradually took effect. Nor did Gerhardt resent correction: he was, on the contrary, proud of Laurent's interest, and studied his precious letters 'like a gospel'.[85]

One of Laurent's most brilliant insights concerned the crucial problem of classification. His chemical types were related by common structures, but, since these structures remained unknown, their relation was of limited use for classification. Gerhardt's proposed classification, based on a vast 'chain of being' whose steps represented different stages of composition and decomposition, was also too general to be useful; every compound could be allotted its place—and then? Laurent did not pretend to understand nature's plans, which *might* involve a chain of being. He was, however, sure that the purpose of a classification was to indicate and establish relations; and this purpose was unfulfilled by Gerhardt's system of 1844.

Without a dominant idea, it is impossible to do anything. Do you ever get anything from your classification? No, nothing, absolutely nothing, because there is no idea there. A classification should offer a series of relations. And I am persuaded that, whatever the point of departure, one always arrives at interesting connections. But still one must start out from an idea.[86]

Laurent perceived that Gerhardt's principle of homology was such an idea, capable of furnishing a basis for relations that could be expressed by suitable formulae.[87] The combination of Laurent's type theory,

[82] *Correspondance de Charles Gerhardt*, vol. 1, p. 278.
[83] Laurent: quoted in Grimaux and Gerhardt, op. cit. (64), p. 105.
[84] Laurent to Gerhardt, 4 July 1844: quoted in Tiffeneau, op. cit. (81), vol. 1, p. 5.
[85] ibid., vol. 2, p. 38: Gerhardt to A.-T. Cahours, 29 May 1845.
[86] Laurent to Gerhardt, 12 Feb. 1845: quoted in Grimaux and Gerhardt, op. cit. (64), p. 474.
[87] ibid., pp. 474–5: Laurent to Gerhardt, 12 Feb. and 24 Feb. 1845.

founded on crystallographic analogies, with Gerhardt's principle of homology, could prove very fertile. Gerhardt defined a series as being composed of a number of molecular systems containing common elements and capable of being metamorphosed into one another:[88] for example, the terms of the paraffin series (ethyl, methyl, etc.) were related by the principle of homology. Each term in the series could be regarded as the 'parent' of a number of derivatives (ethyl alcohol, ethyl chloride, ethyl bromide, etc.), which could be transformed into one another without losing their relation to the parent term. Now Laurent's nucleus theory provided hypothetical aid for envisaging all such transformations, complementing the principle of homology.

Furthermore, if either was to reveal relations, formulae expressing brute chemical composition were inadequate. The formulae of the dualists, at the other extreme, said too much about structure, and Laurent advocated a new approach:

I employ *synoptic formulae*, by the aid of which, I endeavour to manifest certain numerical relations presented by seriated bodies, and to give to analogous bodies analogous formulae. Thus, if I accorded to the sulphates this formula,

$$SM^2 + O^4,$$

and if I considered sulphatic ether as a sulphate, I should represent it also by

$$SEt^2 + O^4,$$

but if, on the contrary, I regarded it as a diamide, I should give to it the formula of the diamides. Be the formula that I employ what it may, for example that of the diamides, it might be considered as a hypothesis; for it is not demonstrated to the satisfaction of every one, that this ether really is a diamide.

But on attentive examination it will be seen, that we have not to ascertain, whether in this ether there is such or such an atomic arrangement, but simply to determine, whether it has the properties of a salt or of a diamide. Here then, we are in the land of experiment.[89]

Laurent's formulae were *not* intended to indicate particular atomic arrangements—only the dualists presumed so far with their formulae. Laurent was content to construct formulae indicating analogies, and his structural ideas were merely used to suggest a theory of combination that would make sense of organic reactions, thereby facilitating chemical classification.

Laurent and Gerhardt were agreed in rejecting the 'rational' formulae of the dualists, and Laurent urged his friend to join him in constructing synoptic formulae 'indicating relations of class and of properties. The others will look for formulae denoting arrangement, and we shall

[88] Gerhardt, op. cit. (80), pp. 228 ff.   [89] Laurent, op. cit. (50), pp. 65–6, 68.

be able to attack them, at least until they find the *true formula*. But they are not there yet.'[90] In the same year, 1845, he repeated his conviction that chemical substances owed their properties largely to their arrangement—which implied that formulae classifying bodies according to their properties would classify them also according to their internal arrangement. 'We do not know what this grouping is. Shall we always be ignorant? Who knows?'[91] Laurent was convinced that the ideal goal of chemical research was a classification that would show how chemical properties were related to molecular arrangement.[92] His conviction that arrangement, although unknown, was influential over chemical properties, emerged repeatedly in his publications.[93] Arrangement was not the source of atomic properties, but rather the factor that modified these properties so as to determine the final properties of the molecular aggregate. For example, in 1846 he discussed reactions such as those between hydrogen and bromine, and between metals and hydrochloric acid, arguing from one-volume formulae.[94] 'This volume, for simple bodies, represents one molecule or two atoms.' (Note his reliance upon the Ampère–Avogadro hypothesis.) Thus oxygen, hydrogen, water (HH)O, phosphine (PH)(HH), etc., could be described as binary compounds. This, of course, did not imply the acceptance of electrochemical dualism; it was meaningless to ascribe a polar structure to the diatomic hydrogen molecule. Laurent suggested that this binary association of atoms perhaps allowed one to explain, to some extent, the affinity possessed by bodies in their nascent state. Let B, B' be bromine atoms, and H, H' hydrogen atoms. Mix BB' with HH', and the affinity of B for B' will perhaps be enough, when joined to that of H for H', to prevent combination; but if only atomic B and H are present, then there is no affinity to be overcome, and combination occurs readily.[95] Thus individual atomic properties were significant, but were subordinated to arrangement and modes of combination. In 1849 Laurent stressed that the idea of different substances playing the same role was 'the most important part of my system, and the one which the chemists will have the most difficulty in admitting'.[96]

Laurent frequently used hypotheses—and this is just the point. He *used* them, but tried to make his *classification* independent of them. 'Such of them as are to be met with [in my work]', he wrote towards the end of his life, 'are isolated, and may be left entirely on one side, without

[90] Laurent to Gerhardt, 19 April 1845: quoted in Grimaux and Gerhardt, op. cit. (64), p. 483.
[91] ibid., p. 487: Laurent to Gerhardt, 11 May 1845.
[92] ibid., p. 495: Laurent to Gerhardt, 29 May 1845.
[93] *C. r. Acad. Sci., Paris* **20**, 357–66 (1845); ibid., **20**, 850–5.
[94] *Ann. Chim. Phys.* (3) **18**, 266–98 (1846).
[95] cf. Laurent, op. cit. (50), pp. 69–70.   [96] *Institut de France* MS. 2379.

any detriment to the progress of the work.'[97] He defined the idea of causality in chemistry as that of atoms and of their arrangement,[98] and added, in Galilean vein, 'I do not reject research after causes, although these may form perhaps but a perpetual mirage, destined to impel us incessantly to an exploration of new countries.'[99] He did not expect answers to his questions about causes, but he asked them, because the search for answers was fruitful. Ultimate answers, however, were going to remain hidden, and Laurent was sure that the important part of his contribution to chemical science was his classification.[100]

Gerhardt, in his next book, published in 1848, adopted and adapted a number of Laurent's suggestions.[101] They agreed that electrochemical dualism was unsatisfactory, being useless for prediction. Gerhardt now asserted that the twin principles of substitution, generally based on double decomposition, and homology, did put one in a position to make predictions, so that knowledge of a single term in a homologous series sufficed for the formal *a priori* deduction of the history of the other terms.[102]

An absolute classification of elements was, he claimed, impossible, because 'one and the same element can entirely change its activity, according to the proportions in which it is combined, and according to the nature of the bodies with which it is associated'. Thus any possibility of a simple, fixed order of affinities was out of the question. Molecular movement presented another bar to dualist ideas about preformation, because it was unreasonable to claim that bodies in motion would maintain the same equilibrium as they enjoyed when at rest.[103] Such arguments might suggest that reactions were no guide to arrangement, just as they were inadequate as a guide to affinity. But, said Gerhardt, if we knew all the possible metamorphoses of a given body, then we should be able to deduce its molecular order.

This is quite impracticable today; we still have a long way to go before we can achieve such progress. Moreover, besides chemical metamorphoses, it is also necessary to study the physical characteristics of bodies, and in particular the relations which exist between their composition and their crystalline form . . . The crystallographic works, with which Messrs. Mitscherlich, Laurent, H. Kopp have already enriched the science, make us feel that the solution of these questions will not be long delayed.[104]

Laurent would have agreed wholeheartedly with this view; many analogies should be considered, especially those between chemical reactions and crystallography. Gerhardt hoped that the search for analogies and relations would one day reveal the laws of chemistry:

[97] Laurent, op. cit. (50), p. xv.   [98] ibid., p. xv.   [99] ibid., p. xvi.
[100] ibid., p. 34.   [101] Gerhardt, op. cit. (80).   [102] ibid., pp. 296 ff.
[103] ibid., pp. 15, 51. cf. Baudrimont.   [104] ibid., p. 55.

. . . we have a deep conviction that all the metamorphoses of a single substance are controlled by general laws, which we can scarcely envisage today, but which the united efforts of chemists will certainly succeed in discovering. . . .

We have been taken to task, with a sort of disdain, for performing chemical algebra; we are glad to accept this description, because we believe that the true progress of science does not consist of limitless multiplications of facts and experiments, but in establishing analogies, and generalising them by formulae, thus finding the laws which are the only guides to the certain prediction of phenomena.[105]

This stress upon the predictive value of general formal laws sounds like good Comteian positivism, and so it would have been, had Gerhardt restricted his analogies to the algebraic formalism of homologous series. But his ideas of analogy extended much further; caution, not positivism, held him back. For the present, formal analogies were the only usefully predictive ones: but the goal of chemistry was to be its unification with physics by the development of sound material analogies.[106]

In 1852 Gerhardt began a series of papers on organic acid anhydrides, which continued into 1853.[107] He argued that it was impossible to formulate double acid anhydrides, such as benzoic acetate, according to dualistic beliefs as combinations of acid with base. Substances were not absolutely electropositive or electronegative: 'There is no absolute opposition between acidic and basic properties; if a single substance can exhibit both, there is obviously only a relative opposition between them; merely a question of degree.'[108] It was therefore ludicrous to try to arrive at a dualistic formulation of the composition of acetic anhydride, benzoic acetate, or compounds such as antimony oxide, which could display both acidic and basic behaviour. Instead, all substances should be arranged in continuous series, analogous to the homologous series whose existence he had demonstrated in organic chemistry.[109] 'To arrange compounds in series, i.e. to determine the laws by which the properties of a given type are modified by substitution of an element or group of elements for other elements, is the constant goal of the philosophical

---

[105] Gerhardt and Chancel, *C. r. mens.* **7**, 65–84 (1851). A. Comte, *Cours de philosophie positive*, vol. 1, pp. 11–14. 1st edn, Paris (1830), 5th edn (1892).

[106] C. Gerhardt, *Traité de chimie organique*, vol. 4, p. 849. 4 vols., Paris (1853–6). 'One can conceive of bodies containing the same atoms combined in the same proportion and arranged in the same manner, that is to say isomeric and isomorphous bodies. However, these bodies need not be identical.' (cf. Laurent's isomeromorphism.)

[107] *C. r. Acad. Sci., Paris* **34**, 755 ff., 902 (1852); he formulated the acid anhydrides according to the water type:

$$\left.\begin{matrix} H \\ H \end{matrix}\right\}O \qquad \left.\begin{matrix} C^2H^5 \\ H \end{matrix}\right\}O \qquad \left.\begin{matrix} C^2H^3O \\ H \end{matrix}\right\}O \qquad \left.\begin{matrix} C^2H^3O \\ C^2H^3O \end{matrix}\right\}O.$$

[108] cf. Gerhardt, op. cit. (69), vol. 2, pp. 495–6.

[109] *Ann. Chim. Phys.* **37**, 285 ff. (1853).

chemist.' Gerhardt proposed four types—water, hydrogen, hydrogen chloride, and ammonia—as the basis for all chemical classification. The admission that there could also be an electrochemical arrangement of the elements was rendered harmless by his insistence that such an arrangement would be relative, not absolute; it would be useless for prediction.

The introduction of four inorganic types was based on the typical relations he had discovered in organic chemistry.[110] Under the guise of what could be misrepresented as the reintroduction of inorganic dualism as the regulative principle for the whole of chemistry, Gerhardt was in fact reinforcing the direction of analogy application pioneered by Dumas, Laurent, and himself. Here, more strongly than was usual with him, he countered the methodological value of any approach based on inorganic ideas about chemical affinity. Even structural analogies, which in this phase of the development of organic chemistry sometimes seemed to bid fair to replace arguments about affinity, have been pushed into the background.

Laurent had long been guided by structural and crystallographic analogies. They were open to abuse, of course, and he, like Dumas, was careful to distinguish between theories and analogies on the one hand and truth on the other. His very reliance upon analogies had enabled him to reprove Gerhardt for allowing structural ideas to creep in unperceived, and, although he modified Gerhardt's approach, Laurent merely fluctuated about a relatively fixed level of confidence in the usefulness of crystallographic analogies. He was rather more attached to them at the beginning and end of his career than in its middle years.

In 1854, Laurent's only extended account of his ideas received posthumous publication, under the title of *Méthode de chimie*. In the following year an English version appeared, translated and introduced by William Odling, who believed 'the generalities of Laurent to be in our day as important as those of Lavoisier were to his'.[111] Laurent had introduced his book as an endeavour to establish a method, 'that is to say, a system of formulae, a classification and a nomenclature, having the advantages of systems based upon facts, *and* of those based upon hypotheses, but without their disadvantages'.[112] The hypotheses in the book (and there are many, dealing with molecular atoms, substitution mechanisms, and crystallographic hints) are mostly separable from the classification.

Towards the end of the book, Laurent came out with an explicit statement of the relation between atomic properties and structural factors:

I admit with all chemists, *that the properties of compound bodies depend upon the nature, the number, and the arrangement of atoms.* But I admit,

---

[110] ibid.   [111] Laurent, op. cit. (50), p. viii.   [112] ibid., p. xv.

moreover, *that order or arrangement has frequently a greater influence upon the properties of the body, than has the nature of the material of which it is composed.*[113]

After a detailed exposition of a series of reactions undergone by aniline and its derivatives, he concluded that mere consideration of affinities, independent of atomic arrangement, led to incorrect conclusions about chemical change. Laurent found this scarcely surprising: 'What, moreover, do we know with regard to affinity?'[114]

Throughout this book, as through the rest of his work, Laurent constantly stressed the fact that arrangements of atoms, although unknown, were directly correlated with properties and reactions.[115] The cardinal sin of electrochemical dualism was that it attempted to ascribe a unique formulation and structure to each substance.[116]

Laurent had written to Gerhardt in 1845 that it was impossible to ascribe unique structural formulae to compounds, not because they did not have a determinate structure, but because 'It seems to me impossible, for the present, to represent a three-dimensional atomic arrangement by a linear formula.'[117] Most chemists at that time, being little concerned with crystallography, were unaccustomed even to thinking of chemical structure in three dimensions—their preoccupation seems to have been more with the order in which atoms were joined, like links in a chain, not bricks in a house; Laurent, aware of these and other difficulties, used his synoptic formulae. Such formulae were amenable to arrangement in series; but it is clear that Laurent intended his formulae to relate to more than formal analogies of composition:

by following out the system of volumes we obtain *the formulae which afford the greatest degree of simplicity; which accord best with the boiling point and isomorphism; which allow the metamorphoses to be explained in the most simple manner, etc.; and, in a word, satisfy completely the requirements of chemists.*[118]

In other words, although the only thing that he absolutely required from his formulae, and the only thing that he would confidently infer from them, was analogy of composition, he still believed that these analogies were related to all other analogies between the various chemical and physical properties of bodies, and that chemists should bear in mind this overall unity of their subject in framing their laws and classifications. Analogies suggested by physical properties, especially those

[113] Laurent, op. cit. (50), pp. 321–2.
[114] ibid., p. 344.   [115] e.g. ibid., p. 268.
[116] Or rather, different dualists were equally vigorous in advancing their own 'unique formulae; see discussion in Laurent, op. cit. (50), pp. 17–33.
[117] Laurent to Gerhardt, 25 March 1845: quoted in Grimaux and Gerhardt, op. cit. (64), p. 481.
[118] Laurent, op. cit. (50), p. 72.

connected with crystallography, were, after the formal analogies of composition, those with which he made the most play.[119] Structural arguments were thus dominant.

Laurent's guiding idea, that crystallography was somehow intimately concerned with chemistry, was eventually to prove fruitful for the whole of chemistry. The problems of structure and classification provided Laurent with a life-time's research, and continued to occupy his successors for many years.

In so far as Laurent's ideas were adopted, the problem of affinity fell into relative neglect. One great difference between Laurent and Berzelius had been that the former sought to understand chemical reactions ultimately in terms of atomic arrangements, while the latter was primarily concerned with atomic properties, affinity chief among them. Yet the theory of types and electrochemical dualism both underwent much modification and elaboration; what might have appeared in 1845 as a fundamental difference of opinion about the importance of affinity could not be viewed so unambiguously by 1855. Types and dualism drifted insensibly towards one another, so that one should ask to what extent the debate between proponents of these seemingly opposed theories expressed real differences in fundamentals. Methodological differences were important; they may have been the only significant ones.

In the following section, these developments and issues will be traced, in relation to chemical affinity.

The unification of chemistry, and its consequences for theories of affinity

Berzelius and Laurent approached chemical experimentation from diametrically opposed positions; but Berzelius himself argued that the very worst way of studying science was to refuse to admit that a given phenomenon could be envisaged in two ways.[120] The development of organic chemistry illustrates the interrelation and eventual reconciliation by mutual modification of the dualist and the unitary approach.

Confirmation of both the radical theory supported by the dualists, and the theories of substitution and of types put forward by the unitarians, came at roughly the same time. The discovery of the persistence

[119] Laurent's interest in crystallography appears in frequent references in papers and on pp. 129–56 of his *Chemical method*. He also compiled for students a *Précis de cristallographie, suivi d'une méthode simple d'analyse au chalumeau* (Paris, 1847), although he was ashamed of the book, which he wrote only for the money. (Tiffeneau, op. cit. (81), vol. 1, pp. 227–8.)

[120] Söderbaum, op. cit. (59), suppl. 2, p. 7 (Stockholm, 1941): letter to A. Marcet, 19 Oct. 1820.

of the benzoyl radical,[121] and the isolation of cacodyl,[122] competed with the discovery of chloracetic acid[123] and of chlorine substitutions into naphthalene and its derivatives. The immediate result was that all parties were strengthened in their conviction that they were right—and their opponents absurdly wrong. Laurent's theory, for example, seemed so extraordinary to Berzelius that he at first accorded it but scant and disdainful treatment in his annual reports.[124] Not all chemists, however, followed the polarization of their fellows into opposed camps. 'Types', the nucleus theory, and electrochemical dualism did not exhaust the available theoretical approaches.

In 1840 Baudrimont published a long article in Quesneville's *Revue*,[125] proposing powerful arguments for the unification of the various branches of chemistry with one another and with physics. Much of the paper was devoted to attacks on Dumas, and to indignant reclamations of priority against him. Baudrimont, unlike Gerhardt and Dumas, believed that laws should unite two orders of phenomena, 'one of which may be considered as a cause and the other as an effect';[126] mere empirical correlation of facts was inadequate.

The so-called law of substitutions was, then, no law at all, although substitutions did exist, both in organic and in mineral chemistry, and were similar in both realms. In spite of this similarity, Dumas had unwisely accepted no help from inorganic phenomena. Baudrimont's criticism implied that the analogy between organic and inorganic chemistry was valid in either direction and that only when this was recognized would one be able to achieve the unification of chemistry. Baudrimont was able to speak in these terms because his corpuscular theory made 'radical' and 'molecule' synonymous. Granted this, it became clear that organic and inorganic chemistry were subject to the same laws; the corpuscular theory provided a model and foundation for the interrelation of chemical and physical properties. Baudrimont had long believed that only evidence from outside chemistry, for example from crystallography or physics, could settle the dualist–unitary debate. Had his approach been followed to its logical end, chemistry would have

[121] F. Wöhler and J. Liebig, *Ann. Phys.* **26**, 325, 465 (1832): trans. in *Am. J. Sci. Arts* **26**, 261–85 (1834): repr. in O. T. Benfey, ed. *Classics in the Theory of Chemical Combination*, pp. 15 ff. Dover, New York (1963).

[122] See Partington, op. cit. (59), vol. 4, pp. 283–6, for list of references and for the significance of cacodyl for the radical theory.

[123] Dumas, *C. r. Acad. Sci., Paris* **7**, 474 (1838); **8**, 609–22 (1839). 1839 was also the year in which cacodyl was first recognized as powerful evidence for the radical theory.

[124] *Jahresber. Fortschr. Chem.* **17**, 225 (1838), after brief summary of Laurent's theory, comments (p. 226), 'Für eine Theorie von dieser Beschaffenheit halte ich eine weitere Berichterstattung für überflüssig.'

[125] *Rev. sci. ind.* **1**, 5–60 (1840).

[126] ibid., p. 32.

vanished as an independent science, and chemical affinity would have ceased to feature as a specific force, as he had already hinted. Chemistry would, in fact, have been absorbed beneath the umbrella of molecular physics. Something of the sort was eventually to happen, but few of Baudrimont's contemporaries heeded his plea for unification, and then only the unification within chemistry.[127]

Bunsen agreed that the same laws held in both organic and inorganic chemistry: chemical affinity and its mode of action were the same in both. If there was any difference between organic and inorganic compounds, this must lie in the nature of organic radicals. Cacodylic acid, for example, was stable towards boiling nitric acid, so perhaps the elements of radicals might somehow be shielded from attack by the forces of affinity.[128] Bunsen regarded affinity as being a function of mass, aggregation, heat, and electricity,[129] but in spite of this dependence affinity 'is in every particular case a certain definitive, unalterable quantity, which like all other forces (and matter itself) can neither be created nor destroyed'.[130] Its mode of action, that is the law regulating the force of affinity, was entirely unknown; the elucidation of this law, although probably far distant, was the most important problem of chemical science.[131]

Concern with the problem of affinity was most active among the dualists, for whom atomic properties held the answer to chemical questions. Nevertheless, some chemists accepted the fact of substitution while still adhering to ideas about the electrochemical properties of bodies. Kolbe and Hofmann were outstanding in this group.

In 1845 Hofmann announced his discovery of organic bases containing chlorine; his researches on the metamorphoses of indigo, aniline, and related compounds, had convinced him 'that in certain circumstances chlorine or bromine can perform the part of hydrogen in organic compounds'.[132] But to some extent the substituent halogen retained its electronegative character, which, as it accumulated, was gradually impressed on the resulting compound. Thus aniline could unite with an acid, bromaniline was less basic, dibromaniline was still less basic, while tribromaniline was neutral.[133] Hofmann had been working on organic bases since the early 1840s, and by 1850 had arrived at the formulation

[127] Baudrimont, *Introduction à l'étude de la chimie par la théorie atomique*, pp. 50, 51, Paris (1833); *Traité de chimie générale et expérimentale*, vol. 1, pp. 210 ff. 2 vols., Paris (1844–6).

[128] *Ann. Chem. Pharm.* **46**, 344 (1843).

[129] *Ann. Chim. Phys.* (3) **38**, 344 (1853).

[130] Bunsen and H. E. Roscoe, *Rep. Br. Ass. Advmt Sci.* 65 (1856).

[131] Bunsen and Roscoe, *Phil. Trans. R. Soc.* **147**, 381 (1857): in part II of their photochemical researches.

[132] *Phil. Mag.* (3) **26**, 515 ff. (1845).

[133] ibid.

of an ammonia type, into which elements or radicals could be substituted.[134] His approach displays an interest in affinity deriving from an interest in individual atomic properties, as well as structural aspects from the theory of types. Like Bunsen, he approached chemistry from both ends of the organic–inorganic axis, and therefore naturally assumed that the same laws of affinity held in both.[135] In later years his work on the phosphorus bases led him to conclude that the phosphine type was precisely parallel to the ammonia type, thereby strengthening the link joining organic and inorganic chemistry.[136] Gradually, however, structural ideas, strongly suggesting typical formulae, became dominant in his mind. When, in 1865, he lectured at the Royal Institution on the combining power of atoms, no mention was made of chemical forces.[137] He illustrated combining power by ball and stick models to illustrate structural formulae, governed by what were in effect valence rules. Combining power now meant no more for him than combining capacity, determined by the number of available valence bonds. In the 1840s he had not gone so far; nor did all those who read of his discovery of bases containing chlorine. Kolbe, for instance, regarded them, and chloracetic acid, as 'conjoined' compounds.[138]

Kolbe argued in favour of Berzelius's methodological rule that 'the application of what is, and will still become known with respect to the mode of combination of the elements in inorganic nature, serves as a guide for the proper judgement of their combinations in organic nature';[139] he was chary of doing homage to the theory of substitutions;[140] and he was a strong supporter of the radical theory.[141] And yet his electrochemical dualism, seemingly one with Berzelius's, was not so very different from contemporary versions of the theory of types. He admitted that reciprocal exchange could take place between hydrogen and chlorine—his own electrolyses had converted chloracetic acid into

[134] *Phil. Trans. R. Soc.* **140**, 93–131 (1850). Wurtz regarded derivatives of the ammonia type as 'important in a theoretical point of view, forming a transition between inorganic and organic chemistry'. (*Phil. Mag.* (3) **34**, 316 (1849)).

[135] *Phil. Trans. R. Soc.* **140**, 93–131 (1850); **141**, 357–98 (1851).

[136] Hofmann and Cahours, *Phil. Trans. R. Soc.* **147**, 575–99 (1857).

[137] *Chem. News, Lond.* **12**, 166, 175, 187 (1865). The papers of T. S. Hunt in the *Am. J. Sci. Arts* illustrated nicely a similar transition, from electrochemical dualism to an acceptance of Laurent's and Gerhardt's theory of types. However, Hunt adhered throughout to a dynamical view of matter, not an atomistic one. The development of his ideas would well repay further study.

[138] *Mem. chem. Soc.* **2**, 360 (1845). Conjoined/conjugated/copulated compounds consisted of two parts in combination; however, modification of one part did not lead to modification of the chemical properties of the other. This explained why chlorine substitution into acetic acid did not greatly affect its properties. cf. Kolbe, *Mem. chem. Soc.* **3**, 378–80 (1848); and note (59) above.

[139] *Trans. chem. Soc.* **4**, 74 (1852).

[140] *Mem. chem. Soc.* **2**, 366 (1845).

[141] *Trans. chem. Soc.* **4**, 73 (1852).

acetic acid.[142] He regarded radicals as 'complex atoms, in which certain atoms may be substituted by others'.[143] He even gave up the notion of a fixed order of electrochemical affinities, supposing that atoms could exist in states in which their electrochemical properties could to some extent be alienated.[144]

Kolbe, in the language of electrochemical dualism, was advocating ideas not far removed from Gerhardt's, and at almost the same time.[145] He came, indeed, to accept the fundamental idea of types, and, as Wurtz remarked angrily in 1860, was so thorough in his acceptance as to feel limited by Gerhardt's mere four types, seeking instead to create yet more types.[146] Kolbe, utilizing both typical and electrochemical concepts in harmony with one another, demonstrated how close together they had become in the years 1845–60.

Frankland was another chemist who combined typical formulae with electrochemical concepts; methyl, ethyl, and homologous radicals 'possess exactly the chemical relations and characters of hydrogen, than which they are, however, less electronegative', and they could replace it wherever it acted as a simple radical. Frankland used typical formulae for organic and inorganic substances,[147] for example,

$$\left.\begin{array}{l} H \\ H \\ H \end{array}\right\}P \qquad \left.\begin{array}{l} C_2H_3 \\ C_2H_3 \\ C_2H_3 \end{array}\right\}P$$

In 1852 he published a paper 'On a New Series of Organic Bodies containing Metals'.[148] He had prepared C_4H_5SnBr, which he called stanethylium bromide. Stanethylium, he found, was stable and persistent through a series of reactions, and seemed in this respect to be analogous to cacodyl. The analogy pointed to a weakness in the theory of conjugation:

It is generally admitted, that when a body becomes conjugated, its essential chemical character is not altered by the presence of the conjunct: . . . therefore, if we assume the organo-metallic bodies above to be metals conjugated with various hydrocarbons, we might reasonably expect, that the chemical relations of the metals to oxygen, chlorine, sulphur, etc. would remain unchanged,

but this was clearly not true. To resolve the anomaly, Frankland proposed a solution arising from consideration of inorganic formulae. There

[142] op. cit. (140).   [143] op. cit. (141).
[144] Paper of 1850: cited by Partington, op. cit. (59), vol. 4, p. 516.
[145] The ideas in the preceding paragraph all date from the years 1845–52.
[146] A. Wurtz, Rép. Chim. pure 2, 349 ff. (1860).
[147] Q. J. chem. Soc. 3, 30 (1851). (Quotation is from p. 51.)
[148] Phil. Trans. R. Soc. 142, 417–44 (1852).

was a clear symmetry between the formulae of many stable compounds with well-satisfied affinities. Consider, for example, the series

$$NH_3 \qquad PH_3 \qquad AsH_3 \qquad PCl_3 \qquad \text{etc.}$$

To account for this symmetry, Frankland suggested that 'no matter what the character of the uniting atoms may be, the combining power of the attracting element, if I may be allowed the term, is always satisfied by the same number of these atoms'.[149] He considered that Laurent and Dumas should have perceived this in their enunciation of the theory of types—but they had gone too far in assuming that the properties of bodies were functions of position and not of the nature of single atoms. Now that the phenomena of substitution were better known, the electrochemical theory could again resume its sway.

The formation and examination of the organo-metallic bodies promise to assist in effecting a fusion of the two theories which have so long divided the opinions of chemists, and which have too hastily been considered irreconcileable; for, whilst it is evident that certain types of series of compounds exist, it is equally clear that the nature of the body derived from the original type is essentially dependent upon the electro-chemical character of its single atoms, and not merely upon the relative position of those atoms.

Frankland's belief that chemistry would develop by means of the synthesis of the theory of types with electrochemical dualism was reinforced by further researches on organometallics. Zincethyl, C_4H_5Zn, reacted violently with the halogens and with sulphur.[150] The violence of reaction supported the views put forward by Brodie,[151] who, at that date, regarded chemical combination as a polar exertion of the forces of chemical affinity in concerted operation brought about by the mutual polarization of reactant molecules, as in the reaction of AgCl with KO:

$$\overset{+}{Ag}\ \overset{-}{Cl}\ \overset{+}{K}\ \overset{-}{O} = AgO + KCl.$$

The reactions of zincethyl also supported Frankland's views

relative to the moleculo-symmetrical form of the organo-metallic compounds. In the inorganic combinations of zinc this metal unites with one atom only of other elements, a very unstable peroxide, not hitherto isolated, being the

---

[149] The full originality of Frankland's insight emerges only when one remembers that the law of multiple proportions had previously led chemists to try to find all the members of series of compounds AB, AB_2, AB_3 etc., and not to look for parallels between formally similar series relating to different sets of compounds:

$$
\begin{array}{llll}
AB & AC & AD & BC & \ldots \\
AB_2 & AC_2 & AD_2 & BC_2 & \ldots \\
AB_3 & AC_3 & AD_3 & BC_3 & \ldots
\end{array}
$$

Vertical series had been sought, to the exclusion of horizontal analogies.

[150] op. cit. (148), **145**, 259–75 (1855).

[151] ibid., **140**, 759–804 (1850).

only exception. The atom of zinc appears therefore to have only one point of attraction, and hence, notwithstanding the intense affinities of its compound with ethyl, any union with a second body is necessarily attended by the expulsion of ethyl.[152]

(It should be noted that Frankland's use of typical formulae was limited to the indication of symmetry and analogy between molecular structures; even when he used Crum Brown's graphic formulae, he insisted 'that such formulae are not meant to indicate the physical, but merely the chemical position of the atoms'.)[153]

In 1850 Williamson published his theory of etherification, schematically expressed as occurring in two stages,

$$\frac{\begin{matrix} H \\ H \end{matrix} SO^4}{\begin{matrix} C^2H^5 \\ H \end{matrix} O} = \frac{\begin{matrix} H \\ C^2H^5 \end{matrix} SO^4}{\begin{matrix} H \\ H \end{matrix} O},$$

then

$$\frac{\begin{matrix} H \\ C^2H^5 \end{matrix} SO^4}{\begin{matrix} H \\ C^2H^5 \end{matrix} O} = \frac{\begin{matrix} H \\ H \end{matrix} SO^4}{\begin{matrix} C^2H^5 \\ C^2H^5 \end{matrix} O}$$

with reciprocal interchange between H and C^2H^5. Since the exchange was continually reversed, it could not be ascribed to the exertion of superior affinities.[154] Casual appeal to affinities irritated Williamson, who deplored the frequent mistaking of names for explanations. 'It unfortunately often occurs that . . . people deceive themselves with the belief that, for instance, in attributing chemical decompositions to affinity . . ., they explain them.'[155] This was not to deny the existence of forces in chemistry; Williamson, however, did not wish merely to refer to affinity in vague terms, but believed that precise quantification should be attempted.[156] He suggested that a first approach to this might be achieved by measurement of reaction rates and concentrations. Gerhardt's concept of chemistry as the study of change, guided by a theory of types,[157] was united with a modified radical theory,[158] and

[152] op. cit. (150).

[153] Frankland and Duppa, *Phil. Trans. R. Soc.* **156**, 309–59 (1866).

[154] *Alembic Club Repr.* no. 16. (Theories of etherification before Williamson's are indicated in Partington, op. cit. (59), vol. 4, pp. 449–50.)

[155] ibid., p. 18: from lecture given 1851 at the Royal Institution.

[156] *Chem. News, Lond.* **3**, 234–9, 246–7 (1861).

[157] Williamson thought that only one type, the water type, was necessary (*Rep. Br. Ass. Advmt Sci.* p. 54 (1851)), and believed that molecular formulae could indicate what a compound really was—he went so far as to discuss mechanical models like orreries. (*Chem. Gaz.* **9**, 334–9 (1851): repr. in Benfey, op. cit. (121), p. 70.)

[158] cf. Kolbe, *Q. J. chem. Soc.* **4**, 73 (1852).

with the study of overall chemical processes. Williamson, in his approach to the problems of chemical affinity, drew on a wide range of ideas. He wished to further not only the unification of the different branches of chemical theory,[159] but also the unification of chemistry with physics. At one point he defined chemical affinity as the process of atomic motion, and attempted to relate specific heat with heat of reaction and with chemical affinity.

. . . the processes where bodies contain much specific heat, and in which the particles are in a very rapid state of motion, which on combining evolve much heat, and accordingly form compounds containing little motion of particles, are those in which the force is greatest during combination, and . . . the tendency of bodies to react chemically, to combine, I will not say in direct proportion to their specific heat, because that would be a quantitative statement which is not authorised at present; but, at all events, . . . those bodies in which the particles are in the most rapid state of motion are those which have the greatest tendency to react chemically, and *vice-versa*.[160]

We have now encountered several organic chemists who, in the early 1850s, were able to reconcile the radical theory, governed by dualistic concepts of electrochemical affinity, with the theory of types, in which implicit structural concepts supplanted conventional ideas about affinity.[161] In 1853 Liebig wrote to Gerhardt, 'It is very strange that the two theories, formerly quite opposed, are now combined in one which explains all the phenomena in the two senses.'[162] This seeming reconciliation was effected by the substantial modification of both theories. The radical theory was able to survive only because the term 'radical' ceased to indicate a stable and all but immutable analogue of the inorganic elements, and began to refer to something less stable; the 'new' radical permitted substitutions to take place,[163] and its properties were

---

[159] *Chem. Gaz.* **9**, 334–9 (1851): repr. in Benfey, op. cit. (121), pp. 69–75.
[160] ibid.
[161] cf. (113) above. C. Daubeny, in *Introduction to the Atomic Theory*, pp. 167–218 (2nd edn, Oxford, 1850), regarded affinity as an inherent atomic property, but considered that the affinity of aggregates depended upon the molecular arrangement. But in his Presidential address to the British Association (*Rep. Br. Ass. Advmt Sci.*, 136 ff.) for 1856, he acknowledged that 'Organic chemistry has also considerably modified our views with respect to chemical affinity', and inclined more towards the theory of types at the expense of dualism.   [162] Quoted by Partington, op. cit. (59), vol 4, p. 460.
[163] cf. Kolbe, op. cit. (158): all the facts adduced by the opponents of the radical theory 'may be made to accord with the above theory, as soon as the idea of the immutability of these radicals is set aside, and exchanged for the opinion that they are complex atoms, in which certain atoms may be substituted by others'. Wurtz (*Rép. Chim. pure* **1**, 24 (1858–9)) also insisted that the radical theory was needed by organic chemistry. 'Que serait la chimie organique sans elle? Maintenant, que l'on cherche à remonter aux éléments eux-mêmes pour expliquer la constitution et les propriétés des radicaux, et en général des combinaisons qui les contiennent, c'est fort bien.' The same point was made by Kekulé, *Ann. Chem. Pharm.* **106**, 129 (1858)—radicals derived their properties from their atomic constituents, and they were not immutable.

capable of being modified by neighbouring substances in combination.[164] Similarly, Berzelius's comparatively rigid electrochemical dualism, whether conventional or realistic, had yielded to the relative dualism of Gerhardt. The latter, being no longer absolute, could serve reliably neither in a normative nor in a predictive role, but could only be used for *post hoc* explanations. Structural analogies suggested by the theory of types had been recognized as less than the whole truth, and atomic properties, including atomicity, began to reassert themselves in the formulation of chemical theory.[165]

The developments of the 1850s contributed to the gradual erosion of precise ideas about chemical affinity. Dualism had been very clear about the problem—Berzelius at least had a specific model—while the theory of types had largely ignored affinity. By the late 1850s, however, these extremes were rare. Affinity was now often regarded as relevant to chemical processes, but little understood. There was no longer any certainty that chemical affinity was even a single special force. Odling, for example, noted that, in metallic precipitations, there was a definite order of reaction in given circumstances, which might be described in terms of affinity, as long as no *theory* was thereby implied. The problem was complex, for, he wrote:

It seems that the affinities leading to the production of any compound are not absolute, but variable, according to circumstances. . . . But irrespective of any absolute variation in the powers of the different affinities, an alteration in the conditions of the experiment brings other forces into play.[166]

Odling was a disciple of Gerhardt, accepting even his highly modified electrochemical ideas.[167] But he lacked his master's caution in considering the molecular constitution of bodies: 'Now we have a period of temperate reaction', he wrote in 1864,

not recognizing the desired knowledge as unattainable, but only as difficult of attainment. And in this, as in many other instances, we find evidence of the healthier state of mind in which, now more perhaps than ever, the first principles of chemical philosophy are explored. Speculation, indeed, is not less rife and scarcely less esteemed than formerly, but is now seldom or never mistaken for ascertained truth.[168]

[164] cf. note (148) above.

[165] This concept was already present in all but name, and forms part of the history of valency.

[166] *Q. J. chem. Soc.* **9**, 289 (1857).

[167] W. Odling, *A Manual of Chemistry*, p. v (1861). 'The views, of which [this Manual] is an exponent, are based on those originally promulgated by Laurent and Gerhardt in France. They are now adopted by a large section of English chemists, particularly by Professors Williamson, Brodie and Hofmann . . .'

[168] *Rep. Br. Ass. Advmt Sci.* address to chemical section, p. 24 (for 1864).

Speculations about chemical affinity were included in this new approach, while dogmatic claims about the nature of chemical affinity became unfashionable. The problem generally resolved itself into one of order—valency and structure—on the one hand, and of energetics on the other.

Prelude to valence theory, and postscript

In 1858 Archibald Scott Couper published papers on affinity in organic chemistry, distinguishing between affinity of degree and elective affinity.[169] The study of affinity, he asseverated, was 'sufficient for the explanation of all chemical combinates'. His discussion was limited to the consideration of affinity bonds, corresponding to valence bonds. However, he prefaced his account with a brief allusion to the electrochemical properties of atoms, which could mutually modify one another. The inevitable result was a vigorous attack from Wurtz, for whom atomicity and arrangement sufficed.[170] Formal analogies, together with analogies of reaction in typical formulae, could be expressed in typical formulae; Wurtz was unwilling to go further.

As far as Wurtz was concerned, ideas about force were of secondary importance to chemistry;[171] his only concern with affinity was where it related to valency. In the preliminary discourse to his *Dictionnaire de chimie* of 1869 he clearly distinguished atomicity (valency) from affinity. Atomicity was concerned with the deployment of energy to a number of other atoms, while affinity was the chemical force, whose nature was unknown. Although he maintained that affinity was elective, the only quantification he regarded as permissible was its correlation with heats of reaction.[172] Wurtz saw the progress of chemistry in the nineteenth century as involving a gradual lessening of interest in the problem of affinity; and, as he wrote, it was clear to him and to his contemporaries that any theory of affinity would be premature—they could see the extent of the problem.[173] Wurtz's analysis, however, was not quite accurate. True, the theory of types had diverted interest from considerations of force to those of structure and patterns of reaction within organic chemistry; but, as this and the next chapter show, the aftermath of the establishment of the theory of types as a theory applicable to the whole of

---

[169] *Alembic Club Repr.* no. 21.

[170] Wurtz, *Rép. Chim. pure* **1** (1858–9). 'Dans les formules de M. Couper . . . la place de chaque atome se trouve marquée non-seulement par le pouvoir basique des éléments, mais encore par je ne sais quelle attraction électrique ou polaire. C'est trop d'hypothèses, et l'on a tort de nous présenter toutes ces choses comme la loi des prophètes.'

[171] cf. note (70) above.

[172] *Bull. Soc. chim. Fr.* **2**, 247, 252 (1864); *Dictionnaire de chimie pure et appliquée*, vol. 1, pp. lxxx–lxxxi. 3 vols. in 5 (1869–78).

[173] ibid., vol. 1, pp. 77–9.

chemistry was not the annihilation of all interest in chemical affinity, but rather the modification of the concept, and its fragmentation into chemical energetics and chemical structure, the distribution of chemical power and matter respectively.[174] In the process of fragmentation and compartmentalization, neither the theory of types nor electrochemical dualism survived in recognizable form. It would be possible to agree either with Wurtz that the theory of types triumphed,[175] or with Blomstrand that dualism was responsible for the state of chemistry at the end of the 1860s;[176] but both, in fact, were modified and fused.

As Crum Brown told the chemical section of the British Association in 1874,[177] an accurate representation of the situation would be,

That the two theories, the dualistic radical theory and the unitary substitution theory, were both true and both imperfect, that they underwent gradual development, scarcely influenced by each other, until they have come to be almost identical in reference to points where they at one time seemed most opposed.[178]

Concluding his address, he introduced the next (and final) chapter: 'We cannot attain to a real theory of Chemistry until we are able to connect the science by some hypothesis with the general theory of Dynamics.'

[174] See Chapter 7.

[175] ibid.

[176] C. W. Blomstrand (*Die Chemie der Jetztzeit vom Standpunkte der elektrochemischen Auffassung aus Berzelius Lehre entwickelt*. Heidelberg (1869)) regarded the theory of types proposed by Gerhardt as merely a special development of Berzelius's atomic theory, and insisted that specific individual atomic affinities were needed for even qualitative explanations of the paths of chemical reactions.

[177] Crum Brown's ideas about chemical affinity are interesting in themselves, and are the subject of investigations by D. Larder. Brown used and inimitably adapted the tenets of the theory of types, joining to them the idea of atomic attractions, which varied individually in intensity, and which interacted with and were modified by the aggregate of attractions within the molecule. (*Trans. R. Soc. Edinb.* **23,** 714–15 (1864)). He expressed these interactions in terms of potential energy (*Proc. R. Soc. Edinb.* **5,** 329 (1866 for 1864)), and had a deeper intuition about the nature of constitution in chemistry than most of his contemporaries had.

We know, indeed, the 'structure' of a considerable number of substances; that is, we know the *order* in which the atoms of these substances are related to each other, but something more than this is implied in the term 'constitution'. . . . For this involves not only the 'structure', or the arrangement of the equivalents in atoms, and in mutually united pairs, but also what we may call the *potential* of each pair of united equivalents. (*Trans. R. Soc. Edinb.* **35,** 151 (1869)).

Unlike Wurtz, he had no wish to neglect affinity for atomicity.

[178] *Rep. Br. Ass. Advmt Sci.* pp. 46, 49 (1874). To illustrate his argument, he referred to Berzelius's view of acetic and trichloracetic acids as being composed of copulae; this was not so far from the 'modern' view that acetic acid was a compound of the radical carboxyl (half a molecule of oxalic acid) and the radical methyl; Berzelius's copulae had, Brown argued, a real and valuable meaning.

NEWTONIAN CHEMISTRY—MATTER AND AFFINITY IN CHEMICAL MECHANICS AND DYNAMICS

HISTORIANS of chemistry have studied the development of ideas of mass action,[1] valency,[2] and thermodynamics,[3] relating these to functional and descriptive aspects of affinity. Thanks to their researches, this chapter will be brief.

Affinity and attraction—C. L. Berthollet

In the 1801 edition of Murray's *Elements of Chemistry*, chemical affinity was described as one of those attractive forces suggested by Newton's 'Queries' to the *Opticks*. The laws governing the exertion of chemical attraction clearly differed from the law of gravitation. Murray noted, however, that 'Some philosophers . . . from a fondness for generalization, have endeavoured to show, that . . . the attractions of aggregation and of combination are ultimately the same with the attraction of gravitation . . . This question is one which it is scarcely possible to decide with certainty, since we have it not in our power to estimate the force of those modifying circumstances which no doubt exist.'[4]

The scientists to whom Murray referred were, in general, very naïve about the complexity of chemical phenomena.[5] Berthollet was unusual for his time in combining considerable chemical sophistication with Newtonianism. His earlier notions, however, seemed more conventional, distinguishing chemical from mechanical laws, and insisting upon the importance of elective affinities.[6] Admittedly, gravitational attraction

---

[1] F. L. Holmes, 'From elective affinities to chemical equilibria: Berthollet's law of mass action'. *Chymia* **8**, 105–45 (1962).

[2] C. A. Russell, *The History of Valency*. Leicester University Press (1970); W. G. Palmer, *A history of the concept of valency to 1930*. Cambridge (1965).

[3] V. M. Schelar, 'Thermochemistry, thermodynamics and chemical affinity'. *Chymia* **11**, 99–124 (1966). E. Hiebert, *Historical roots of the principle of conservation of energy*. Madison, Wis. (1962).

[4] J. Murray, *Elements of Chemistry*, vol. 1, pp. 65–7. 2 vols., Edinburgh (1801).

[5] e.g. A. Libes, *J.Phys.* **59**, 391 (an 10); Governor Pownall, *Phil. Mag.* **18**, 155 (1804).

[6] e.g. *Ann. Chim. Phys.* **9**, 138 (1791); chemical adhesion cannot be referred to mere mechanical laws; specific chemical properties imply specific affinities, and it is by a succession of these affinities that, in the case of dyestuffs, colouring molecules fix themselves to cloth.

was not merely mechanical; but Berthollet in 1791 identified attraction of adhesion as mechanical, and attraction of composition as chemical. In later years, he ignored this division. In 1801 he announced that:

The theory of chemical affinities solidly established, and serving as a basis for the explanation of all chemical questions, ought to be a collection of, or contain, all the principles from which the causes of chemical phenomena can proceed, in every possible variety of circumstances because observation has proved, that all these phenomena are only the various effects of that affinity, to which all the various chemical powers of bodies may be attributed.[7]

Previous theories of affinity, Bergman's in particular, had regarded the order of affinities as fixed, and chemical reactions as determined by this order. Berthollet was over-simplifying here, presenting a bold target for criticism. He aimed to prove that affinity did not act alone as the determining force in chemical reaction, but that the relative proportions of the reactants were also influential. In addition, other modifying factors, such as insolubility, cohesion, crystallization, heat, and the elasticity of gases, had to be taken into account. The influence of the relative proportions of reactants could be deduced from Berthollet's picture of a chemical system as a static analogue of the solar system—a number of bodies whose forces were in equilibrium;[8] solution, accordingly, was indistinguishable in kind from chemical combination.

Berthollet's acceptance of the Newtonian paradigm was made explicit at the beginning of his *Essai de statique chimique*. He regarded affinity and gravitation as arising probably from one and the same attribute of bodies; but only the former was affected by the shapes of particles, by distances between them, and by elective affections.[9] Substances might possess several such affections, one of which was in some cases dominant, determining overall chemical properties and principal kinds of combination. Inflammable substances, for example, were distinguished by their affinity for oxygen, an affinity between opposites. Similarly, 'the reciprocal affinity of acids and alkalis constitutes acidity and alkalinity'.[10] Berthollet, extending Lavoisier's dualism, identified chemical combination, the mutual satisfaction of affinities, with the neutralization and equilibration of properties and forces respectively.[11] He stressed again the importance of the relative quantities of reactants, citing Newton and Laplace to reinforce his argument.[12]

The authority of Berthollet and of Laplace ensured that, in the early

---

[7] C. L. Berthollet, *Researches into the Laws of Chemical Affinity*, trans. Farrell, p. 1. French edn, Paris (1801), this edn, London (1804).

[8] Attractive, for one another, and repulsive, due to caloric.

[9] Berthollet, op. cit. (7), pp. 1 ff.

[10] C. L. Berthollet, *Essai de statique chimique*, vol. 1, p. 61. 2 vols., Paris (1803).

[11] ibid.   [12] ibid., vol. 1, pp. 522–3, 532–5.

years of the nineteenth century, Newtonianism was the respectable orthodoxy of French science.[13] Their writings, however, were mainly effective in revealing the multiplicity of factors that had to be reckoned with, thereby removing any immediate prospect of predictive certainty from chemical science. Dumas, an admiring student of the *Statique chimique*, nevertheless complained that he had found its exposition confusing and difficult to understand;[14] nor was he alone in this.[15] Laplace, in his *System of the World*, suspected that universal and chemical attraction were ultimately the same, while admitting that such generalizations were premature, in ignorance of the laws of affinity.[16] A major influence of these patriarchs of French science may well have been to divert attention from the frustrating search for a quantitative chemistry based upon tables of affinity, towards the quantitative investigation of chemical proportions. This would also have been promoted by the Berthollet–Proust controversy about the existence of definite proportions; Berthollet's defeat would only have reinforced the trend towards quantitative analysis and the determination of atomic weights. There at least careful work yielded precise results, unlike broad but shallow theorizing.

In 1804 the *Edinburgh Review* summarized Berthollet's ideas on chemical affinity, pointing out that their influence on the study of affinity would be mainly confusing and destructive; the former pillar of chemistry, the fixity of chemical relations, would be toppled.[17] There was, inevitably, a certain reluctance to accept this blow, but criticisms of Berthollet, favourable or adverse, were few and for some years his ideas received little or no development.[18] As Dalton noted, the contemporary view of chemical affinity 'is daily growing more [obscure] in proportion to the new lights attempted to be thrown upon it'.[19] Cuvier,

[13] cf. M. Crosland, *The Society of Arcueil*, pp. 259, 300, 327. London (1967).

[14] J.-B. Dumas, *Leçons sur la philosophie chimique recueillies par M. Bineau*, pp. 379 ff. Paris (1837).

[15] See, for example, note (17) below.

[16] P. S. Laplace, *System of the World*, trans. J. Pond, vol. 2, pp. 235 ff. 2 vols., London (1809).

[17] A. Duncan, *Edinb. Rev.* **5**, 141 ff. (1804).

[18] F. L. Holmes, op. cit. (1), gives a detailed history of the development of chemical statics to their fruition in mass action, and has a full list of references. The early debate, e.g. Schnaubert versus Pfaff versus Berthollet, revolved around the challenge that Berthollet had issued to supporters of affinity tables as useful in chemistry.

Mém Soc. Arcueil (e.g. Gay Lussac and Thenard, **2**, 295 (1809); A. B. Berthollet, **1**, 304 (1807)) provides good examples of chemistry within Berthollet's paradigm; in the former paper, the reasoning was based upon the variability of chemical affinity (in connection with the preparation of alkali metals), while the latter stressed the importance of chemical mass, and referred to the *Statique chimique*.

[19] J. Dalton, *A New System of Chemical Philosophy*, part 1, p. 142. Manchester (1808). Beyond his notion that atoms were surrounded by envelopes of caloric, and that when they combined, the resulting molecules possessed a single envelope of caloric, and

in contrast, gave strong support to Berthollet's ideas on affinity, believing that not only were they true, but also they had the advantage

of binding chemistry more closely to the great system of the physical sciences. On the other hand, the simple consideration of affinity and the tacit exclusion of the ordinary forces of nature seemed to leave [chemistry] in the state of isolation in which its creatures had placed it. The chemist . . . will no longer be able to dispense with being a physicist and geometer.[20]

Newtonianism in chemistry would indeed have bound the science more closely to physics, but at the beginning of the nineteenth century all such attempts were premature. Brewster, an ardent Newtonian, recognized this when he spoke of 'those mysterious relations among the elementary principles of matter which hypothesis has scarcely ventured to anticipate'.[21]

In the first thirty years of the nineteenth century, electrical theories attracted far more attention than conventional Newtonian chemistry, which underwent little development. The primacy of affinity was still recognized, but largely by way of lip-service; it was of no account in the laboratory. For more than a decade after Berthollet's death, affinity theory in France remained sterile and stationary. The revival of French science after Napoleon's downfall was at first totally unconnected with the development of mechanical theories of affinity—perhaps as a revulsion from the very Newtonianism of the Napoleonic heads of science.

In England, meanwhile, dynamical speculations abounded. J. B. Emmett, for example, published a series of 'Researches into the mathematical principles of chemical philosophy',[22] in which he asseverated that the *only* force between particles was inversely proportional to the square of the distance between them. Significantly, he regarded atomic heat capacities and thermochemistry in general as being fundamental to a complete 'mathematico–mechanical' chemical philosophy. At about the same time John Herapath published his first paper on what was to become known as the kinetic theory of gases, in which he attempted to relate together Newtonian attraction, heat, and chemical affinity.[23] Most English chemists, however, were not mathematically sophisticated. Brougham, in asking Mary Somerville to translate Newton's *Principia*

---

granted the reservation that chemical combination was governed by largely geometrical considerations, Dalton was little interested in the mechanism of chemical combination. In a MS. note of a lecture Dalton gave in 1834 (Instn elect. Engrs MS. Dalton, 6 May 1834), the problem of chemical combination is raised only to be dismissed. Considerations of *forces* in chemistry seem scarcely to have troubled John Dalton.

[20] cf. G. Cuvier, *Rapport historique sur les progrès des sciences naturelles depuis 1789*, pp. 24–5. Paris (1810).

[21] *Phil. Mag.* **45**, 118 (1815).

[22] *Ann. Phil.* **9**, 421 (1817), through to *Ann. Phil.* N.S. **10**, 372 (1825).

[23] ibid., N.S. **1**, 273, 340, 401 (1821).

and Laplace's *Mécanique céleste*, bemoaned his countrymen's ignorance of these works;[24] sustained mathematical exertion was welcomed by few; but speculations, explaining all the phenomena of chemistry in general terms of matter, attractive forces, and motion, abounded in the literature.[25] In the third edition of *On the Connexion of the Physical Sciences*, Mary Somerville, influenced by Faraday and serving as a barometer for the climate of scientific opinion, discussed chemical affinity and the law of molecular action at short distances:

Whether it be a modification of gravity, or that some new and unknown power comes into action, does not appear. But as a change in the law of the force takes place at one end of the scale, it is not impossible that gravitation may not remain the same throughout every part of space. Perhaps the day may come, when even gravitation, no longer regarded as an ultimate principle, may be resolved into a yet more general cause, embracing every law that regulates the material world.[26]

Early thermochemistry in relation to chemical affinity

In 1839, G. H. Hess carried out a series of measurements of heats of dilution, which he believed gave a thermochemical measure of affinity.[27] In the following year, Baudrimont re-stated his dynamical theory of chemistry, and urged that the thermal properties of bodies be considered in relation to their chemical reactivity.[28]

The possible relation or interaction between heat and chemical affinity

[24] Bodleian Library, Somerville Papers: letter of 17 March 1827 (Mrs. Somerville's *Mechanism of the Heavens* was published in London in 1831). cf. Playfair, *Edinb. Rev.* (1808): quoted by J. T. Merz, *A History of European thought in the nineteenth century*, vol. 1, p. 232 (4 vols., repr. Dover, New York, 1965): 'We will venture to say that the number of those in this island who can read the *Mécanique céleste* with any tolerable facility is small indeed.' See also *Q. Rev. chem. Soc.* **1**, 108 (1809): 'We do not believe that ten persons in the universe have read Laplace's *Mécanique céleste* as it ought to be read.'

[25] e.g. Marshall Hall, *Nicholson's J.* **30**, 193 (1811)—effects are the result of heterogeneous and homogeneous attraction, which produce a constant order of affinities (cf. Bergman); Sir Richard Phillips, in *Phil. Mag.* **56**, 195 (1820), explained all galvanochemical phenomena in terms of 'the mechanical Theory of Matter and Motion'; J. M. Good (*The Book of Nature*, vol. 1, p. 105. 3 vols., London (1826)) wrote of chemical affinity: 'It is highly probable that this kind of attraction is also nothing more than a peculiar modification of that of gravitation, more select in its range, but more active in its power.' T. Exley (*Principles of Natural Philosophy; or a new theory of physics, founded on gravitation*. London (1829)) included chemical affinity within his gravitational theory; and, as a final instance in a list that could be protracted indefinitely, John James Waterston, in *Phil. Mag.* N.S. **10**, 170 (1831), expounded a 'New Dynamico-Chemical Principle'.

[26] M. Somerville, *On the connexion of the Physical Sciences*, pp. 408–409. 3rd edn, London (1836).

[27] See H. A. M. Snelders, 'De Thermochemische onderzoekingen van Hess (1839–42)'. *Chem. Tech. Rev.* **20**, 397–9 (1965).

[28] *Rev. Sci. Ind.* **1**, 5–60, esp. 34–6 (1840).

had long been considered, at least speculatively. In 1798, Rumford had raised the question 'whether chemical affinity be not the mere effect of temperature'.[29] Perhaps, he suggested, different degrees of heat would produce great differences in specific gravity, which might somehow account for chemical reaction. Not until about 1840, however, did such speculations receive experimental investigation with a view to illuminating the problem of affinity. Lavoisier's pioneering work in thermochemistry had been designed to follow the transformations involving substantial caloric,[30] but only when heat was regarded as a form of force, energy, or motion could one relate it to chemical force.

James Prescott Joule, convinced that the correlation of forces, including heat, was an empirical fact, began to investigate systematically the relations of electricity, heat, and chemical affinity to mechanical power in terms of the work each could perform. He belonged unequivocally to the energetic school. The British Association's *Report*[31] for 1842 announced that Joule's earlier hypotheses, that the heat of combustion had an electrical origin, and 'that the affinity of atoms for one another is the measure of the heat evolved by their combination',[32] had received both theoretical and experimental confirmation. In the latter connection, the agreement of Joule's results with Dulong's gave additional weight to the demonstration.[33]

Joule expressed succinctly his conviction of the interrelation of the powers of nature in 1847:

Indeed the phenomena of nature, whether mechanical, chemical, or vital, consist almost entirely in a continual conversion of attraction through space, living force, and heat into one another. Thus it is that order is maintained in the universe—nothing is deranged, nothing ever lost, but the entire machinery, complicated as it is, works smoothly and harmoniously . . . the whole being governed by the sovereign will of God.[34]

In 1850 Joule published his famous paper 'On the mechanical equivalent of heat',[35] in which he acknowledged his debt to Faraday for the proof of the identity of chemical and electrical forces, and for the idea 'that the so-called imponderable bodies are merely the exponents of different forms of force. Mr. GROVE and M. MAYER have also given their

[29] *Nicholson's J.* **2**, 162 (1798).

[30] Lavoisier had, however, collaborated with Laplace, who thought of heat as arising from motion. Their results were so expressed that it would be possible, although decidedly unhistorical, to interpret them in terms of chemical energetics.

[31] 'On the Electric Origin of the Heat of Combustion'. *Rep. Br. Ass. Advmt Sci.*, p. 31 (for 1842).

[32] *Phil. Mag.* **19**, 275 (1842); **20**, 111 (1843).

[33] *C. r. Acad. Sci., Paris* **7**, 871 (1838).

[34] J. P. Joule, 'On Matter, Living Force and Heat': lecture 1847: repr. in I. B. Cohen and H. M. Jones, ed. *Science before Darwin*, p. 173. London (1963).

[35] *Phil. Trans. R. Soc.* **140**, 61–82 (1850).

powerful advocacy to similar views.' Joule furnished additional experimental support for these views by demonstrating the 'equivalence' of heat and mechanical work. He reaffirmed that 'the heat evolved by the combustion of a body is proportional to the intensity of its affinity for oxygen', and in the next three years related thermochemistry to electrical theories of combination.

It will be sufficient for my purpose to admit,—1st., that when two atoms combine by combustion, a current of electricity passes from the oxygen to the combustible; 2nd., that the quantity of this current of electricity is fixed and definite; and 3rd., that it is the means of evolution of light and heat, precisely as is any other current of electricity whatever.[36]

These assumptions, buttressed by the work of, among others, Becquerel and Faraday,[37, 38] implied that electrical measurements would yield values for heats of combination, and hence for 'the absolute force with which bodies enter into combination'. Electrochemistry and thermochemistry would both yield quantitative measures of affinity.

There were of course others besides Joule who were engaged in thermochemical researches. Thomas Graham, for example, read a paper to the Chemical Society on heats of combination,[39] and Andrews published a series of papers on the heat changes accompanying acid–base reactions and basic substitutions.[40] In a paper of 1848 about heats of combustion, he concluded with the remark that

as the heat developed in chemical reactions may be taken as a measure of the forces brought into play, I deemed it proper to refer to the foregoing cases [combustion with oxygen and chlorine], if only for the purpose of directing attention to the intimate relations which enquiries of this kind have with some of the most interesting questions of molecular chemistry.[41]

Andrews offered the first clause of the passage above without any attempt at explanation or justification; although one could not assume widespread agreement concerning the nature of chemical affinity, one could, apparently, assume that chemical work, the operation of affinities, was in quantitative correlation with heats of combustion. Joule's researches had clearly contributed to this climate of opinion, which was merely one aspect of a general and rising conviction of the correlation

---

[36] J. P. Joule, 'On the heat disengaged in chemical combinations'. *Phil. Mag.* (4) **3**, 481 (1852).

[37] Becquerel, *Traité de l'électricité*, vol. 2, p. 85. 7 vols., Paris (1834–7).

[38] See Chapter 3 above for Faraday's work on the correlation of forces in connection with chemical affinity.

[39] *Mem. chem. Soc.* **1**, 106 (1843).

[40] See esp. *Phil. Trans. R. Soc.* **134**, 21–37 (1844); *Phil. Mag.* (3) **32**, 321 (1848); (3) **36**, 511 (1850)—a report and summary of the state of knowledge of thermochemistry.

[41] *Phil. Mag.* (3) **32**, 434 (1848).

and conservation of physical forces[42]—a conviction existing even before its precise formulation.

Thus Chambers could assert, 'The inorganic has one final comprehensive law, GRAVITATION.'[43] Goodman spoke to the 1848 meeting of the British Association 'On the identity of the existences or forces of light, heat, electricity, magnetism and gravitation',[44] in the following year G. Wilson suggested once again to the Chemical Society that 'mechanical' and chemical forces might differ only in degree and not in kind,[45] and in 1851 William Thompson, proposing a 'mechanical' theory of electrolysis, looked forward to the foundation of a 'mechanical' theory of chemistry.[46]

This exceedingly general picture needed to gain in depth and detail. On the one hand, chemical thermodynamics, concerned with processes rather than with explanatory models, did not yet exist; on the other hand, mere correlation of chemical and physical forces did not really help one to picture the actual mechanism of chemical reactions. To remedy the latter situation, a particularly interesting model was proposed by Williamson, who, in the face of considerable scepticism about atoms, firmly advocated atomic motion as the foundation of any chemical theory.[47] He proposed a dynamical statistical model of chemical equilibrium, in contrast with the chemical statics of an earlier generation, and held to a kinetic theory of heat. Combination was accompanied by heat changes, and by changes in chemical properties. Williamson commented:

It is to be hoped that we may soon be able to give an account of the nature of the processes by which these changes of properties are effected; but that task can only be entered upon when we have obtained exact determinations of the relative momentum of atoms in various compounds, the proportion of which to their masses determines their physical and chemical properties.[48]

Thermochemical investigations multiplied. A very extensive series was carried out by Favre and Silbermann, and published at considerable

[42] Hence, presumably, the successive editions that were required of W. R. Grove, *Correlation of physical force* (6th edn, London, 1874).
[43] R. Chambers, *Vestiges of the natural history of creation*, p. 362. 2nd edn, London (1844).
[44] *Rep. Br. Ass. Advmt Sci.* p. 53 (for 1848).
[45] *Q. J. chem. Soc.* **1**, 174 (1849).
[46] *Phil. Mag.* (4) **2**, 429 (1851). For 'mechanical' (their term) read 'dynamical'.
[47] *Rep. Br. Ass. Advmt Sci.* **2**, 65 (for 1850); *Phil. Mag.* **37**, 350 (1850); *C. r. Acad. Sci., Paris* **16**, 23. 'In using the atomic theory, chemists have added to it of late years an unsafe, and, as I think, an unwarrantable hypothesis, namely that the atoms are in a state of rest. Now this hypothesis I discard, and reason upon the broader basis of atomic motion.' (*Phil. Mag.* **37**, 356 (1850).)
[48] *Chem. Gaz.* **9**, 334–9 (1851).

length in 1852 and 1853.[49] They had carried out hundreds of exact calorific measurements of chemical changes because 'Everything which concerns the laws of chemical affinity, and the molecular modifications undergone by bodies which change their chemical properties without any change in weight, is still enveloped in deep enough obscurity.'[50] They described experiments with a water calorimeter to determine various heats of combination, then went on to consider the calorimetry of chemical segregation (dry decomposition) and of reactions in solution, including acid–base neutralization, double decomposition, and dissolution. From this last class of reactions, they constructed tables of 'calorific equivalents', defined as the quantities of heat evolved by equivalent weights in combination. They found that the more stable compounds were those whose formation had been most exothermic. Their discussion of calorific equivalents necessarily referred to systems in comparable physical states, since no significant results could be obtained by comparing, for example, the thermochemistry of gas reactions with that of reactions between solids. They took reaction in solution as their standard, and proceeded to correlate equivalent heats of formation with the 'energy of chemical affinities'. The latter was undefined, but seems to have been based on a combination of stability with behaviour in displacement reactions. The authors wrote hundreds of pages summarizing the results of hundreds of experiments, all to illuminate a concept that they refused to define. There is at once an element of circularity and of cautious probing in their work. Starting with affinity as a thoroughly vague concept, they set out to correlate it with thermochemical properties. Gradually, by implication rather than by explicit statement, the idea of affinity as somehow expressing the aggregate of stability, calorimetrically determined energy of formation, and relative position in a displacement or electrolysis series emerged from the fact that this nebulous aggregate could be correlated with a specific value—the 'calorific equivalent'. Once this correlation had been found, affinity was specifically identified with the correlate of calorific equivalents, which, in the cases they investigated, turned out to be chemical stability. In their view, the constituents of a stable compound had powerful affinities for one another. By the end of their researches, Favre and Silbermann, having chosen this interpretation, were able to talk unhesitatingly about the energy of affinities. They measured the heats of combination of different bases for the same acid, expressed their results in terms of calorific equivalents, tabulated them, and commented:

[49] *Ann. Chim. Phys.* (3) **34**, 357 (1852); **36**, 5 (1853); **37**, 406 (1853); **37**, 484 (1853).
[50] ibid., **34**, 357 (1852).

Glancing at these numbers, it seems difficult not to admit that there is quite a close connection between the energy of the affinities of different bases for the same acid, or the degree of stability of the compound formed, and the quantities of heat evolved in the act of combination.[51]

The thermochemical interpretation of chemical affinity was gaining in precision, and thermochemistry continued to attract the best efforts of leading scientists. Kopp, Hess, Graham, Andrews, Thomsen—these and many more vigorously pursued the relations between heat and chemistry. In 1856 Becquerel added plausibility to their view that affinity was correlated with heats of combustion.[52] He relied on 'the ordinary laws of the evolution of heat by electricity' to confirm and extend the relation. Briefly, he argued that, as a consequence of Faraday's laws of electrolysis, one could state that the decomposition of equivalents of bodies evolved equal amounts of electricity. A simple calculation then sufficed to show that the heat arising from chemical action was proportional to the electromotive force produced by the reaction. Heats of reaction calculated on the basis of this conclusion were in moderate agreement with those determined experimentally by Favre and Silbermann.

This convenient dovetailing of results, and indeed the whole trend of thermochemistry, only confirmed Favre in his belief that calorimetry was the chemists' indispensable tool. 'To study chemical reactions, taking into account the quantities of heat involved, is, in our opinion, the best, perhaps the *only* way by which one may arrive at a correct conception of the force designated by the name of affinity.'[53, 54]

Marcellin Berthelot and chemical mechanics

Among the most influential thermochemical investigations around 1860 were those of Marcellin Berthelot. Berthelot's prestige was such that a special chair in chemistry was created for him at the *Collège de France*, and the recommendation for his appointment was couched in terms implying that there was no other suitable candidate.[55] His own propaganda had created the misleading impression that before him there had been no such thing as organic synthesis,[56] the field wherein he published

[51] ibid., **37**, 491 (1853).   [52] *Ann. Chim. Phys.* (3) **48**, 257 (1856).

[53] P. A. Favre and C. du Queylar, *C. r. Acad. Sci., Paris* **50**, 1150 (1860).

[54] Granted this approach, endothermic reactions posed a problem, for they seemed to violate the principle that the reciprocal attraction of heterogeneous molecules produced heat. Favre suggested that simultaneous diffusion occurred, lowering the temperature. If this was a large enough effect, then the net result would be the absorption of heat. (*C. r. Acad. Sci., Paris* **51**, 316 (1860).)

[55] *Académie des Sciences*, Paris MS.: letter of Wurtz to Dumas, 30 Dec. 1863, which describes the affair with high indignation.

[56] See esp. his *Chimie organique fondée sur la synthèse chimique.* 2 vols., Paris (1860). It is summarized in his *Science et philosophie*, pp. 41 ff. Paris (1886).

his first papers. In the course of these synthetic researches, however, he evinced a growing interest in reaction mechanisms, and in the problem of chemical affinity. In a paper of 1856 on the relations between carbonic oxide and formic acid, he noted that cuprous chloride solution absorbed carbon monoxide, forming an unstable crystalline hydrate.[57] The instability of this and of related compounds was accounted for by the speed of reaction, which was so great as to lead to little alteration in 'molecular state' between the reactants and products. In cases where the reaction was slower, latent affinities could come into play, and, through slow and extensive changes, could lead to the formation of more stable compounds.

Berthelot pursued his researches into the correlation between reaction rates and affinity, publishing his first series of results in 1861.[58] He co-operated with Péan de St.-Gilles in a study of the formation and decomposition of ethers, according to the reversible reaction

$$2\text{ROH} \;\overset{\text{(acid)}}{\underset{}{\rightleftarrows}}\; \text{ROR} + \text{H}_2\text{O}.$$

They tested the effects of different acids, varying proportions, dilution, and pressure. Their initial conclusions merely related the effects of dilution to alterations in intermolecular distances. They considered that there were two conflicting factors: (i) increase in intermolecular distance led to a decrease in reaction rate, and (ii) the greater separation of molecules allowed the combination between acid and alcohol to reach its maximum proportion.

Subsequent researches emphasized the role played by time and equilibrium conditions in what was now described as a 'whole new chemical mechanics'.[59, 60] Equilibrium became their key concept. They found that etherification was reversible, so that the reaction between acid and alcohol never reached completion, but instead tended towards an equilibrium state depending upon the relative numbers of equivalents of acid and alcohol present. Furthermore, the reversible nature of the reaction made it irrelevant to the final equilibrium whether equivalents of 'products' or of 'reactants' were originally present. They clearly attached much importance to the notion of reversibility, yet they did not consider it in formulating a mathematical expression for

[57] *Ann. Chim. Phys.* (3) **46**, 477 (1856).

[58] *C. r. Acad. Sci., Paris* **53**, 474 (1861); **55**, 39; **55**, 210, 324: *Ann. Chim. Phys.* **65**, 385 (1862): reviewed in *Rép. Chim. pure* **4**, 1, 325, 327, 369 (1862).

[59] Berthelot and St.-Gilles, *Ann. Chim. Phys.* (3) **65**, 385–422 (1862); **66**, 5–110 (1862); **66**, 110–28 (1862); **68**, 225 (1863); Berthelot, *Ann. Chim. Phys.* (4) **6**, 290 (1865).

[60] Around this date several other chemists were carrying out observations on rates of reaction, e.g. J. H. Gladstone, *Phil. Trans. R. Soc.* **145**, 179 (1855); *Phil Mag.* **9**, 535 (1855); *J. chem. Soc.* **9**, 54 (1856); **15**, 302 (1862). See also papers by L. F. Wilhelmy, esp. in *Ann. Phys.* **81**, 413, 499 (1850), on the inversion of cane sugar.

reaction velocity, and therefore failed to arrive at the law of mass action.[61]

Berthelot and St.-Gilles noted that the proportion of ether formed in different systems was independent both of the nature of the mineral acid used in etherification, and of the particular alcohol used. This had important consequences for the old idea of specific and elective affinities:

The ideas of particular and individual affinities, which were previously regarded as playing such an important part in etherification, must henceforth give way to a very simple and very general notion, founded essentially upon equivalents. The statics of etherification reactions thus return into the same fundamental law as the statics of mineral reactions.[62]

The next stage in the development of Berthelot's concept of affinity was the identification of chemical mechanics with general mechanics, and the consequent attribution of heats of reaction to molecular impact.[63] Heats of reaction thus appeared as functions of only the initial and final mechanical states of chemical systems.[64] Within this scheme, chemical affinity took on a rather insignificant role; it was responsible for initiating any given chemical transformation, but the latter always occurred strictly according to the laws of mechanics. The forces of chemical affinity were the only ones to which Berthelot appealed, but they seem functionally analogous to, and as sterile as, God in a wholly mechanical universe; once the system has been set in motion, it will operate according to well-defined laws, which have no further dependence upon their originator.

In his *Essai de mécanique chimique fondée sur la thermochimie*,[65] Berthelot gave an extended and definitive account of his theory of chemical reaction, supported by many calorimetric results. He defined affinity as the resultant of the actions that held the elements of compound bodies together.[66] One could measure the work done by affinity by the heat changes accompanying reaction.[67] Chemical reaction

[61] For the mathematical expression, as also for references to the work of Guldberg and Waage (who formulated the law of mass action), see J. R. Partington, *A History of Chemistry*, vol. 4, pp. 584 ff. London (1961–4).

[62] Berthelot and St.-Gilles, *Ann. Chim. Phys.* **68**, 358 (1863).

[63] ibid., (4) **6**, 292 (1865).

[64] ibid., 294. This is in fact a paraphrase of Hess's law of heat summation (1841), which had served as a correction to Dulong's erroneous conclusion of 1839 (cf. *Ann. Chim. Phys.* **8**, 180) that the heat of combustion of a compound is equal to the sum of the heats of combustion of its elements. Favre and Silbermann had made use of Hess's law in their work (see note (77) below). Berthelot considered that one could deduce the principle *a priori* from the notion that heats of reaction arose from molecular impact. (*Ann. Chim. Phys.* (4) **6**, 294 (1865)).

[65] M. Berthelot, *Essai de mécanique chimique fondée sur la thermochimie*. 2 vols., Paris (1879).

[66] ibid., vol. 1, p. xxiv.

[67] ibid., vol. 2, pp. xxviii–xxix.

A M—P

occurred in accordance with the principle of maximum work, tending to produce the body or system of bodies whose formation evolved the most heat.[68] Berthelot's theory of chemical combination was entirely compatible with classical mechanics. Heat, a form of motion, became manifest when bodies lost kinetic energy. A precise understanding of chemical combination would require a knowledge of the mass, kinetic energies, and individual motions of all the particles in the system.[69] Such knowledge was impossible, so chemists had to rely upon the overall information given by heats of reaction.

Berthelot explained endothermic reactions by postulating either an external supply of energy—for example, the electric arc lamp provided energy for the endothermic formation of acetylene—or the occurrence of simultaneous exothermic reactions.[70] The latter also accounted for the old baffling notion of 'predisposing affinities',[71] which can best be explained by an example. Carbon monoxide will not combine directly with water, but these substances will yield formic acid in the presence of a caustic alkali. The latter could be said to have brought into play a predisposing affinity, causing reaction between carbon monoxide and water, which makes possible the subsequent favoured reaction between formic acid and caustic alkali. There was, however, something mysterious about this sequence. The reaction (i), $CO + H_2O = HCOOH$, would not occur, unless it could be followed by, say, (ii), $KOH + HCOOH = H_2O + HCOOK$. But how did (i) know that it would be followed by (ii)? If, however, one supposed with Berthelot that (i) and (ii) were simultaneous, and that (ii) evolved more heat than (i) absorbed, then the combined reactions in the presence of an alkali would be exothermic, and so would readily occur.

Berthelot wished to explain the whole of chemistry, at least in principle, by two opposing factors: the inertia and consequent stability of molecular types, and the principle of maximum work. Heats of reaction gave the measure of molecular forces, in accordance with the laws of rational mechanics:[72] 'From the mechanical point of view, . . . the mass of the elementary particles, . . . and the nature of their motions, . . . should be sufficient to explain everything.'[73] This was the fundamental result of the new science. The reciprocal actions of bodies could be foreseen with certainty as soon as their individual 'conditions of existence' were known. Thermochemistry thus represented merely a stage in the progress of chemistry towards its inclusion in 'the unity of the universal law of . . . natural *forces*', a law applying equally to stars

[68] M. Berthelot, *Essai de mécanique chimique fondée sur la thermochimie*, vol. 1, p. 13, 2 vols., Paris (1879).

[69] ibid., vol. 2, p. 2.   [70] ibid., vol. 2, pp. 18–29.   [71] ibid., vol. 2, p. 31.

[72] M. Berthelot, *Science et philosophie*, pp. 100, 103. Paris (1886).

[73] ibid., p. 102.

and to atoms.[74] Somehow Berthelot was able to persuade himself that the mechanical basis of thermochemistry arose 'not from obscure and uncertain glimpses, not following *a priori* reasoning, but by means of indubitable notions, always founded on observation and experiment'. Thermochemistry, indeed, appeared to him as a rich fruit of 'positive science'.[75]

Berthelot's attempted reduction of chemistry to a science of bodies characterized essentially only by their mass and motion represents a high point of nineteenth-century classical mechanism. It was bold, clear, and, as later developments were to show, fundamentally unsound. J. B. Stallo, acute critic of atomo-mechanical physics, had a field day with the *Mécanique chimique*, whose very title recalled the work and aims of Laplace.[76]

Prelude to thermodynamics

The Atomic Debates of the London Chemical Society showed that belief in chemical atomism was far from universal:[77] nevertheless, the rise of the kinetic theory in physics and the growth of physical chemistry both helped to produce a climate favourable to atomism.[78] Thomas Anderson, for example, discussed Brodie's chemical calculus before the British Association, remarking that 'the real subject of inquiry is . . . the unit of matter'. The molecular hypothesis was necessary in physics, 'and it is scarcely possible to doubt that some connection must exist between the chemical and physical unit of matter'.[79] James Clerk Maxwell also was surprised that chemists should doubt the atomic theory when physicists accepted the kinetic theory of gases.[80] Similar examples could be multiplied readily,[81] mechanical theories similar to Berthelot's being not uncommon in the early 1860s.[82] This approach

[74] ibid.

[75] ibid., p. 9 (reprint of lecture of 1863).

[76] J. B. Stallo, *Concepts and theories of modern physics*, ed. P. Bridgman, pp. 306 ff. 3rd edn (1888), repr. Cambridge, Mass. (1960).

[77] See W. Brock, *The Atomic Debates*. Leicester (1967).

[78] Introductory address to the chemical section of the British Association: reported in *The Laboratory*, **1**, 397–400 (1867).

[79] ibid.

[80] *Chem. News, Lond.* **15**, 303 (1867).

[81] e.g. Thomas Graham, a physical chemist, clearly adhered to a molecular hypothesis (*Phil. Trans. R. Soc.* **153**, 385 ff. (1863); *Phil. Mag.* (4) **27**, 81 (1864)). Williamson's adherence to the atomic theory has received frequent mention.

[82] Mechanical theories were proposed by Maumené, *Bull. Soc. chim. Fr.* **2**, 129 (1864); E. Czirnianski, *Theorie der chemischen Verbindungen, auf der rotierenden Bewegung der Atome basiert* (Krakau, 1863): summarized in *Jahresber. Fortschr. Chem.* p. 89 (1863); Sainte-Claire Deville, *Phil. Mag.* (4) **32**, 365 (1866); A. Dupré, *Ann. Chim Phys.* (4) **13**, 64 (1868): who hoped to persuade theoreticians that chemical phenomena were the immediate consequences of the general law of universal attraction.

In his Presidential Address to the British Association in 1866 (*Rep. Br. Ass. Advmt*

however, was soon found to have shortcomings. Mechanics alone did not suffice for the total explanation of chemical qualities and reactions, so that affinity was again invoked to take care of awkward properties. Whether it was a distinct force or the resultant of a number of concurring factors, affinity had some relation to the chemical nature of the bodies concerned.[83] Equivalents were not enough in considering chemical reactions and equilibria; they could not account for variable atomicity, nor for the specificity of reactions, except when the mechanical theory was phrased so generally as to be liable to the Popperian criticism of explaining everything and therefore really explaining nothing.

In 1866 A. Dupré, in his fifth memoir on the mechanical theory of heat, announced that he had found 'that the law of attraction at small distances is not the same for different substances, or at least that the constants included in its expression depend upon the chemical nature of the bodies'.[84] In the same year Chevreul insisted that there was such a thing as elective affinity, which was a force opposing and distinct from cohesion and gravity.[85] Even those, like H. Sainte-Claire Deville, who, believed that affinity could not escape 'the identification of the forces which come into play in the physical and chemical phenomena of nature',[86] were not happy with the existing state of the concept. They considered that affinity was retained as a blanket term for phenomena

Sci. p. lxi) W. R. Grove stated that the correlation of forces, including chemical affinity, was now proved, so that 'in chemical inquiries, as in other branches of science, we are gradually relieving ourselves of hypothetical existences', including, he implied, the existence of a distinct force of chemical affinity; affinity, with all other forces, resolved itself objectively into motion.

In 1871 P. G. Tait told the British Association (Rep. Br. Ass. Advmt Sci. p. 5) that there was little doubt that the principle of the conservation of energy 'contains implicitly the whole theory of Thermo-electricity, of Chemical Combination' etc., etc.; and in 1876 Thomas Andrews again suggested to the British Association (Rep. Br. Ass. Advmt Sci. p. lxvii), that the division between physics and chemistry was probably artificial.

Two others who proposed mechanical theories of chemistry were Friedrich Mohr, who, in his Mechanischen Theorie der chemischen Affinitaet (Braunschweig, 1868), offered a thermochemical explanation of theoretical chemistry; and A. Naumann, who observed (Grundriss der Thermochemie, p. 150 (Braunschweig, 1869)), that 'chemistry in its ultimate form must be atomic mechanics'.

[83] For two objections that more than mechanics was involved in chemical phenomena see Crum Brown, Trans. R. Soc. Edinb. 23, 714–16 (1864), where different atoms were attributed different intensities and distribution of affinity within a wider-than-classical mechanical scheme; and A. Naquet, The Laboratory 1, 327–30 (1867), who insisted that affinity was more than an ordinary physical force, since it produced the phenomena of definite combining proportions.

[84] Ann. Chim. Phys. (4) 9, 329 (1866).

[85] M. E. Chevreul, Histoire des connaissances chimiques, vol. 1 (vol. 2 was never published), p. 43. Paris (1866).

[86] Phil. Mag. (4) 32, 365 (1886). Sainte-Claire Deville's work on dissociation was a pioneering investigation of chemical process, a valuable step towards chemical thermodynamics.

that were not understood; it should either be explained in terms of universal physics, or else rejected as occult. Rejection was straight-forward, requiring merely the explication of chemical combination in terms of recognized natural forces. Deville tried to achieve this.[87]

Those who wished to retain the term 'affinity' were faced, as ever, with its definition. Berthelot's thermochemical answer, that heats of combination gave a measure of affinities, was incomplete. Some com-pounds, after all, were endothermic yet remarkably stable: surely high stability could not arise from the exertion of negative affinities.[88] A model-free solution to the problem was eventually forthcoming from the second law of thermodynamics, where affinity appeared as free energy, the maximum external work of reversible isothermal reactions. A complementary avenue of development came from the growth of valence theory, through the discovery of the electron and the constitu-tion of elemental atoms to contemporary quantum-mechanical inter-pretations of electronic theories of valency and of chemical bonding. Such developments, however, lie well beyond the confines of this book, which has been concerned to trace developments of nineteenth-century affinity theory only to their pre-modern stage.

[87] *The Laboratory* **1**, 63–5 (1867).
[88] Schroeder van der Kolk, *Ann. Chim. Phys.* (4) **4**, 193–208 (1865).

BIBLIOGRAPHY

PRIMARY SOURCES

(A) MANUSCRIPT SOURCES

The following libraries have provided manuscript sources.

Académie des Sciences, Paris.
Bodleian Library, Oxford.
British Museum, London.
Cambridge University Library.
Hollandsche Maatschappij der Wetenschappen, Haarlem.
Institut de France, Paris.
Institution of Electrical Engineers, London.
Museum of the History of Science, Oxford.
Magdalen College, Oxford.
Royal Institution, London.
Royal Society, London.
Trinity College, Cambridge

(B) PRINTED SOURCES

1. Journals and Periodicals used

Alembic Club Reprints.
American Journal of Science (ed. Silliman).
Annalen der Physik.
Annales de chimie.
Annales de chimie et de physique.
Annals of Philosophy.
Bulletin de la Société Chimique de France.
The Chemical Gazette.
The Chemical News.
Compte rendu hebdomadaire des séances de l'Académie des Sciences.
Compte rendu (mensuel) des travaux chimiques (ed. Gerhardt).
Edinburgh Journal of Science.
Edinburgh New Philosophical Journal.
Edinburgh Review.
Gentleman's Magazine.
Jahresbericht über die Fortschritte der Chemie.
Jahresbericht über die Fortschritte der physische Wissenschaften.
Journal of the Chemical Society.
Journal de pharmacie et des sciences accessoires.
Journal of Natural Philosophy, Chemistry, and the Arts (ed. Nicholson).
Journal de physique, de chimie, et d'histoire naturelle.
Journal of the Royal Institution.

The Laboratory.
Mémoires de l'Académie des Sciences.
Mémoires de physique et de chimie de la Société d'Arcueil.
Nature.
The Philosophical Magazine.
Philosophical Transactions of the Royal Society.
Proceedings of the Royal Society.
Proceedings of the Royal Society of Edinburgh.
Quarterly Journal of Science, Literature, and Arts.
The Quarterly Review.
Répertoire de chimie pure.
Répertoire de chimie appliquée.
Report of the British Association for the Advancement of Science.
Revue scientifique et industrielle (ed. Quesneville).
Scientific Memoirs (ed. Taylor).
Transactions of the Royal Society of Edinburgh.

2. Selected list of books used

AMPÈRE, A. M. *Correspondance du grand Ampère*, ed. L. de Launay. 3 vols., Paris (1936–43).

BACON, F. *Works.* 2 vols., London (1838).

BAUDRIMONT, A. E. *Introduction à l'étude de la chimie par la théorie atomique.* Paris (1833).

—— *Rev. sci. ind.* **1,** 5–60 (1840).

—— *Traité de chimie générale et expérimentale.* 2 vols., Paris (1844–6).

BEDDOES, T. *Contributions to Physical and Medical Knowledge, principally from the West of England.* Bristol (1799).

BENCE-JONES, H. *The Life and Letters of Faraday.* 2 vols., London (1870).

BENFEY, O. T., ed. *Classics in the Theory of Chemical Combination.* Dover, New York (1963).

BERGMAN, T. O. *Physical and Chemical Essays*, vols. 1 and 2, trans. E. Cullen. 3 vols. London (1784–91).

—— *A Dissertation on Elective Attractions*, trans. from 1775 Swedish edn. London (1785).

BERTHELOT, M. *Chimie organique fondée sur la synthèse.* 2 vols., Paris (1860).

—— *Essai de mécanique chimique fondée sur la thermochimie.* 2 vols., Paris (1879).

—— *Science et philosophie.* Paris (1886).

BERTHOLLET, C. L. *Researches into the Laws of Chemical Affinity*, trans. Farrell from 1801 French edn. London (1804).

—— *Essai de statique chimique.* 2 vols., Paris (1803).

BERZELIUS, J. J. *An attempt to establish a pure scientific System of Mineralogy by the application of the Electro-chemical Theory and the Chemical Proportions*, trans. John Black. (1841).

—— *Lärbok i Kemien*, translations:
  Elemente der Chemie der unorganschen Natur, trans from 1st Swedish edn. Leipzig (1816).

Lehrbuch der Chemie, trans. from 2nd Swedish edn (1820–8); 3rd edn, trans Wöhler, 10 vols., Dresden and Leipzig (1833–41); 4th edn, 10 vols., Dresden and Leipzig (1835–41); 5th edn, 5 vols., Dresden and Leipzig (1843–8).

Traité de chimie, trans Jourdain and Esslinger from 1st Swedish edn. 8 vols., Paris (1829–33).

—— *Essai sur la théorie des proportions chimiques, et sur l'influence chimique de l'électricité*, trans. from Swedish. Paris (1819); 2nd edn, Paris (1835).

—— *Jahres-Bericht über die Fortschritte der physischen Wissenschaften*, 20 vols. (1822–41).

—— *Jahres-Bericht über die Fortschritte der Chemie und Mineralogie*, vols. 21–7. 7 vols. (1842–8).

—— *Jac. Berzelius Bref.*, ed. Söderbaum. 6 vols. + 3 suppl., Uppsala (1912–32).

BLACK, J. *Lectures on Chemistry*, ed. Robison. 2 vols., Edinburgh (1803).

BLOMSTRAND, C. W. *Die Chemie der Jetztzeit vom Standpunkte der electro-chemische Auffassung aus Berzelius' Lehre entwickelt.* Heidelberg (1869).

BOERHAAVE, H. *Elementa chemiae.* 2 vols., Fritsch, Leipzig (1732).

BOSCOVICH, R. J. *Theoria philosophiae naturalis* (Venice 1763 edn), trans. J. M. Child. Chicago (1922).

BOSTOCK, J. *An Account of the History and Present State of Galvanism.* London (1818).

BOYLE, R. *The Origine of Formes and Qualities.* London (1667).

—— *Works of the Honourable Robert Boyle*, ed. T. Birch. 6 vols., London (1772).

BRANDE, W. T. *A Manual of Chemistry.* London (1819); 4th edn (1836).

BRODIE, SIR B. C. *The Works of Sir B. C. Brodie.* 3 vols., London (1865).

BYKOV, G. V., ed. *Centenary of the Theory of Chemical Structure.* Moscow (1961). Reprints of valuable and inaccessible papers.

CHAMBERS, R. *Vestiges of the Natural History of Creation.* 2nd edn, London (1844).

CHEVREUL, M. E. *Histoire des connaissances chimiques.* Paris (1866).

COLERIDGE, S. T. *Collected Letters of S. T. Coleridge*, ed. Griggs. 4 vols., Oxford (1956–9).

—— *The Notebooks of Samuel Taylor Coleridge*, ed. K. Coburn. London (1957–).

COMTE, I. A. M. F. X. *Cours de philosophie positive.* 6 vols., Paris (1830–42).

—— *The Positive Philosophy*, trans. H. Martineau. 2 vols., London (1853).

CUVIER, G. *Rapport historique sur les progrès des sciences naturelles depuis 1789.* Paris (1810).

—— *Histoire des progrès des sciences naturelles.* Paris (1834).

DALTON, J. *A New System of Chemical Philosophy.* Manchester (part 1, 1808; part 2, 1810; part 1 of vol. 2, 1827).

DANIELL, J. F. *Introduction to Chemical Philosophy.* 2nd edn (1843).

DAUBENY, C. G. B. *Introduction to the Atomic Theory.* 2nd edn, Oxford (1850).

—— *Miscellanies.* London and Oxford (1867).

DAVY, H. 'An Essay on Heat, Light and the Combinations of Light' in *Contributions to Physical and Medical Knowledge, principally from the West of England*, ed. T. Beddoes. Bristol (1799).

Davy, H. *Researches Chemical and Philosophical, chiefly concerning Nitrous Oxide*. (1800).

—— *Elements of Chemical Philosophy*. London (1812).

—— *Consolations in Travel; or, The Last Days of a Philosopher*. London (1830).

—— *Collected Works*, ed. J. Davy. 9 vols., London (1839).

Davy, J. *Memoirs of the Life of Sir Humphry Davy*. 2 vols., London (1836).

—— *Fragmentary Remains, Literary and Scientific, of Sir H. Davy*. London (1858).

Digby, Sir Kenelm. *The Nature of Bodies; and the Nature of Man's Soule*. London (1645).

—— *Of the Sympathetick Powder*. London (1669).

Donovan, M. *Essay on Galvanism*. Dublin (1816).

—— *Chemistry*. 4th edn, London (1839).

Dumas, J.-B. *Leçons sur la philosophie chimique, recueillies par M. Bineau*. Paris (1837).

Edgeworth, M. *Life and Letters of Maria Edgeworth*, ed. A. J. C. Hare. 2 vols., London (1894).

Encyclopaedia Britannica. 2nd edn, Edinburgh (1778); 3rd edn (1797); 4th edn (1810); 5th edn (1817).

Exley, T. *Principles of Natural Philosophy*. London (1829).

Faraday, M. *Experimental Researches in Electricity. Reprinted from the Philosophical Transactions of 1831–1852*. London (1839–55).

—— *The Subject Matter of a Course of Six Lectures on the Non-Metallic Elements*, ed. J. Scoffern. London (1853).

—— *Experimental Researches in Chemistry and Physics*. London (1859).

—— *A Course of Six Lectures on the Various Forces of Matter and their Relations to each other*, ed. W. Crookes. London (1860).

—— ['Some Thoughts on the conservation of Forces' in *The Correlation and Conservation of Forces: a Series of Expositions, by Prof. Grove, Prof. Helmholtz, Dr. Mayer, Dr. Faraday, Prof. Liebig, and Dr. Carpenter. With an Introduction and Brief Biographical Notices of the Chief Promoters of the New Views*, ed. E. L. Youmans. New York (1865).] I have been unable to find a copy of this work.

—— *The letters of Faraday and Schoenbein*, ed. G. W. A. Kahlbaum and F. V. Darbishire. London (1899).

—— *Faraday's Diary (being the various philosophical notes of experimental investigation made by Michael Faraday during the years 1820–1862 and bequeathed by him to the Royal Institution of Great Britain)*, ed. T. Martin. 7 vols. + index, London (1932–6).

Fourcroy, A. F. *Elements of Chemistry, and Natural History, to which is prefixed the Philosophy of Chemistry*, trans. R. Heron from 4th French edn. 4 vols., London (1796).

Fownes, G. *A Manual of Elementary Chemistry*. London (1844); 6th edn. (1856); 11th edn (1873); 13th edn (1883).

Freind, J. *Praelectiones Chymicae*. London (1809).

—— *Chymical Lectures: in which almost all the Operations of Chymistry are Reduced to their True Principles, and the Laws of Nature*. London (1712).

GERHARDT, C. *Précis de chimie organique.* 2 vols., Paris (1844–5).
—— *Introduction a l'étude de la chimie par le système unitaire.* Paris (1848).
—— *Traité de chimie organique.* 4 vols., Paris (1853–6).
—— *Correspondance,* ed. Tiffeneau. 2 vols., Paris (1918–25).
GMELIN, L. *Hand-Book of Chemistry,* trans. H. Watts. 19 vols., London (1848–70).
GOOD, J. M. *The Book of Nature.* 3 vols., London (1826).
GRAHAM, T. *Elements of Chemistry.* 2nd edn, 2 vols., London (1850).
—— *Chemical and Physical Researches,* ed. A. Smith. (1876).
GRAVES, R. P. *Life of Sir W. R. Hamilton.* 3 vols., London and Dublin (1882–9).
GREN, F. A. C. *Principles of Modern Chemistry.* 2 vols., London (1800).
GRIMAUX, E. and GERHARDT, CH. JUN. *Charles Gerhardt, sa vie, son œuvre, sa correspondance.* Paris (1900).
GROVE, W. R. *The Correlation of Physical Forces.* 6th edn, London (1874).
HARRINGTON, R. *The Death-warrant of the French Theory of Chemistry, . . . with a theory fully . . . accounting for all the Phenomena. Also a full . . . Investigation of . . . Galvanism.* Carlisle (1804).
—— *An Elucidation and Extension of the Harringtonian System of Chemistry, explaining all the Phenomena; Without one single Anomaly.* London (1819). All Harrington's writings are more stimulating than significant in their eccentricity.
VAN HELMONT, J. B. *Oriatrike.* London (1662).
HENRY, W. *An Epitome of Chemistry.* London (1801).
—— *Memoirs of John Dalton.* London (1854).
HERSCHEL, J. F. W. *Preliminary Discourse on the Study of Natural Philosophy.* London (1830).
—— *Essays from the Edinburgh and Quarterly Reviews.* London (1857).
—— *Popular Lectures on Scientific subjects* (1895).
HIGGINS, W. *A Comparative View of the Phlogistic and Anti-Phlogistic Theories* (1789); repr. Wheeler & Partington, *The Life and Work of William Higgins* (Pergamon Press, 1960).
—— *Experiments and Observations on the Atomic Theory, and Electrical Phenomena.* London (1814).
HOFMANN, A. W. 'The Life-work of Liebig'. *Faraday Lecture for 1875.* London (1876).
—— *Inaugural Address.* Berlin (1880), trans. Boston (1883).
VON HUMBOLDT, A. *Cosmos,* trans. E. C. Otté. 5 vols. (1849–58).
HUNT, T. S. The Relations of the Natural Sciences. *Canadian Naturalist* **10,** no. 5 (1882).
—— *A New Basis for Chemistry, a Chemical Philosophy.* Boston (1887).
JOULE, J. P. 'On Matter, Living Force and Heat' (1847). *Science Before Darwin,* ed. Cohen, I. B. and Jones, H. M. London (1963).
KANE, R. *Elements of Chemistry.* 2nd edn, Dublin (1849).
KANT, I. *Metaphysical Foundations of Science,* trans. B. Bax from 1786 Riga edn. London (1883).
—— *Critik der reinen Vernunft.* Riga (1781).

KNIGHT, G. *An Attempt to Demonstrate, that all the Phaenomena in Nature may be Explained by two simple Principles, Attraction and Repulsion*. London (1748).

KOPP, H. *Geschichte der Chemie*. 4 vols., Brunswick (1843–7); repr. Leipzig (1931).

—— *Die Entwicklung der Chemie in der neuren Zeit*. Munich (1873).

LAPLACE, P. S. *System of the World*, trans. J. Pond. 2 vols., London (1809).

LAURENT, A. *Précis de cristallographie, suivi d'une méthode simple d'analyse au chalumeau, d'après des leçons particulières de M. Laurent*. Paris (1847).

—— *Méthode de chymie*. Paris (1854).

—— *Chemical Method*, trans. W. Odling. London (1855).

LAVOISIER, A. L. *Traité élémentaire de chimie*. 2 vols., Paris (1789).

—— *Elements of Chemistry*, trans. R. Kerr. Edinburgh (1790).

LE SAGE, G. L. *Essai de chymie mécanique*. (1758).

VON LIEBIG, J. and KOPP, H. *Jahresbericht über die Fortschritte der Chemie*. Giessen.

—— *Familiar Letters on Chemistry*. London (1843).

LOW, D. *An Inquiry into the Nature of the Simple Bodies of Chemistry*. London (1844); 2nd edn (1848); 3rd edn (1856).

MACQUER, P. J. *Elements of the Theory and Practice of Chemistry*, trans. A. Reid. 2 vols., London (1775).

—— *A Dictionary of Chemistry*, trans. J. Keir. 3 vols., London (1777).

MAKO, P. *Compendiaria Physicæ Institutio quam in usum auditorum philosophiae elucubratus est*, part 1. 2nd edn, Vindobonæ (1766).

MAXWELL, J. C. *Scientific Papers*, ed. W. D. Niven. 2 vols., Cambridge (1890).

MENDELEEFF, D. *The Principles of Chemistry*, trans. G. Kamensky from 5th Russian edn, ed. A. J. Greenaway. 2 vols., London (1891). The book was written in 1868–70.

MITSCHERLICH, A., ed. *Gesammelte Schriften von Eilhard Mitscherlich*. Berlin (1896).

MOHR, F. *Mechanischen Theorie der chemischen Affinitaet*. Brunswick (1868).

MURRAY, J. *Elements of Chemistry*. 2 vols., Edinburgh (1801).

NAUMANN, A. *Grundriss der Thermochemie*. Brunswick (1869).

NEWTON, I. *Philosophiae Naturalis Principia Mathematica*. London (1687).

—— *Mathematical Principles of Natural Philosophy*, trans. A. Motte, revised F. Cajori. Berkeley (1934).

—— *Opticks*. London (1704); 4th edn (1730), repr. Dover, New York (1952).

—— *Isaac Newton's Papers and Letters on Natural Philosophy*, ed. I. B. Cohen. Cambridge (1958).

—— *Correspondence of Isaac Newton*, ed. H. W. Turnbull. Cambridge (1960).

—— *Unpublished Scientific Papers of Sir Isaac Newton*, ed. A. R. and M. B. Hall. Cambridge (1962).

NICHOLSON, W. *A Dictionary of Chemistry*. 2 vols., London (1795).

NITSCH, F. A. *A General and Introductory View of Professor Kant's Principles concerning Man, the World, and the Deity*. London (1796).

ODLING, W. *A Manual of Chemistry*. (1861).

—— *Outlines of Chemistry*. London (1870).

ODLING, W. *Abstract of Lectures delivered at the Royal Institution*. London (1874).

OERSTED, H. C. *Ansichten der chemischen Naturgesetze* Berlin (1812). The British Museum copy is annotated by S. T. Coleridge.

—— *Recherches sur l'identité des forces chimiques et électriques*, trans. Chevreul. Paris (1813).

—— *The Soul in Nature*. London (1852).

—— *Naturvidenskabelige Skrifter*. 3 vols., Copenhagen (1920).

—— *Correspondance de H. C. Oersted avec divers savants*, ed. Harding. 2 vols., Copenhagen (1926).

OSTWALD, W. *Naturphilosophie*. (1901); 2nd edn, Leipzig (1902).

—— *The Principles of Inorganic Chemistry*, trans. A. Findlay. 3rd edn, London (1908).

—— *Natural Philosophy*, trans. T. Seltzer. London and New York (1911).

PALEY, W. *Natural Theology*. 1st edn, London (1802); with notes by H. Brougham and C. Bell. 2 vols., London (1836).

PARIS, J. A. *The Life of Sir Humphry Davy, Bart*. 2 vols., London (1831).

PARKES, S. *Chemical Catechism*. London (1806); 10th edn (1822).

PRIESTLEY, J. *The History and Present State of Electricity*. London (1767).

—— *Disquisitions on Matter and Spirit*. 2nd edn, 2 vols., London (1782).

RITTER, J. W. *Über das electrische System der Körper*. Leipzig (1805).

—— *Die Begründung der Elektrochemie und Entdeckung der ultravioletten Strahlen*, ed. A. Hermann. Frankfurt (1968).

ROBISON, J. *A System of Mechanical Philosophy*, ed. Brewster. 4 vols., Edinburgh (1822).

ROSCOE, H. E. and HARDEN, A. *A New View of the Origin of Dalton's Atomic Theory*. London (1896).

ROWNING, J. R. *A Compendious System of Natural Philosophy* . . . London (1774).
This was a popular text book, and went through several editions; Bodley has 4 between 1738 and 1759.

VON SCHELLING, F. W. J. *Ueber Faraday's neueste Entdeckung*. Munich (1832).

—— *Sämmtliche Werke*, ed. K. F. A. Schelling. 14 vols., Stuttgart and Augsburg (1856–61).

SOMERVILLE, M. *On the Connexion of the Physical Sciences*, London (1834); 3rd edn (1836); 9th edn (1858).

—— *Molecular and Microscopic Science*. 2 vols., London (1869).

STALLO, J. B. *The Concepts and Theories of Modern Physics* (repr. of 3rd edn 1888), ed. P. Bridgeman. Cambridge, Mass. (1960).

STEWART, D. *Philosophical Essays*. Edinburgh (1810).

STEWART, B. and TAIT, P. G. *The Unseen Universe* (1876).

TAIT, P. G. *Recent Advances in Physical Science*. 3rd edn, London (1885).

THENARD, L. J. *Traité de chimie élémentaire, théorique et pratique*. 4 vols., Paris (1813–16); 4th edn, 5 vols. (1824); 5th edn (1827).

THOMPSON, S. P. *Michael Faraday. His Life and Work*. London (1898).

THOMSON, T. *A System of Chemistry*. 4 vols., Edinburgh (1802); 4th edn, 5 vols., Edinburgh (1810); 5th edn, 4 vols., London (1817).

—— *A System of Chemistry of Inorganic Bodies*. 7th edn, London (1831).

THOMSON, T. *An outline of the sciences of heat and electricity*. London (1830).

—— *The History of Chemistry*. 2 vols., London (1830–1).

—— and TAIT, P. G. *Principles of Mechanics and Dynamics (Treatise on Natural Philosophy)*. 2 vols., repr. Dover, New York (1962).

TURNER, E. *Elements of Chemistry*. Edinburgh (1827). This book went through 8 edns in 20 years.

TYNDALL, J. *Faraday as a Discoverer*. London (1868).

WHEWELL, W. *An Essay on Mineralogical Classification and Nomenclature with Tables of the Orders and Species of Minerals*. Cambridge (1828).

—— *History of the Inductive Sciences*. 3 vols., London (1837).

—— *Philosophy of the Inductive Sciences*. 2 vols., London (1840).

—— *Lectures on Education*. London (1855).

—— *On the Philosophy of Discovery*. London (1860).

WILLIAMSON, A. W. *Chemistry for Students*. Oxford (1865).

WILSON, G. *Chemistry*. Edinburgh (1850).

—— *Religio Chemici*. London and Cambridge (1862).

WURTZ, A. *Leçons de philosophie chimique*. Paris (1864).

—— *Dictionnaire de chimie pure et appliquée*. 3 vols., Paris (1869–78); Suppl. 2 vols. (1880–6).

—— *A History of Chemical Theory from the Age of Lavoisier to the Present Time*, trans. H. Watts. (1869).

—— *La théorie atomique*. (1879).

YOUNG, T. *A Course of Lectures on Natural Philosophy and the Mechanical Arts*. 2 vols., London (1807).

SECONDARY SOURCES

1. Journals and periodicals used

Ambix.

Annals of Science.

Archives for the History of the Exact Sciences.

Archives internationales d'histoire des sciences.

British Journal for the History of Science.

British Journal for the Philosophy of Science.

Centaurus.

Chymia.

Isis.

Journal of Chemical Education.

Revue d'histoire des sciences et de leurs applications.

2. Selected list of printed books used

VON AESCH, A. G. *Natural Science in German Romanticism*. New York (1941)

BENFEY, O. T., ed. *Classics in the Theory of Chemical Combination*. Dover, New York (1963).

BROCK, W. *The Atomic Debates*. Leicester (1967).

CARDWELL, D. S. L., ed. *John Dalton and the Progress of Science*. Manchester and New York (1968).

CLAGETT, M., ed. *Critical Problems in the History of Science*. Madison (1959).
COHEN, I. B. *Franklin and Newton: an Inquiry into Speculative Newtonian Experimental Science and Franklin's Work in Electricity as an Example thereof*. Cambridge, Mass (1966).
—— and JONES, H. M., ed. *Science Before Darwin*. London (1963).
COPLESTON, F. *A History of Philosophy*. London (1946–).
CROSLAND, M. *The Society of Arcueil*. London (1967).
DAS, H. *Over de historische ontwikkeling van het begrip* 'molecuulverbinding'. Ph.D. thesis, University of Amsterdam; publ. Delft (1962).
Dictionary of National Biography. Oxford.
EDWARDS, P., ed. *Encyclopedia of Philosophy*. 8 vols., New York and London (1967).
ELLOWITZ, J. The history of the theories of chemical affinity from Boyle to Berzelius. M.Sc. thesis, University of London (1927).
Encyclopaedia Britannica, 11th edn.
ERIKSSON, G. Berzelius och atomteorin: den idehistorisken bakgrunden. *Lychnos*, 1–37 (1965). (English summary 35–37.)
FISHER, N. Chemistry B.A. Part II thesis, Oxford University (1965).
GREGORY, J. C. *The Scientific Achievements of Sir. H. Davy*. London (1930).
—— *A Short History of Atomism*. London (1931).
HALL, A. R. and M. B. *Unpublished Scientific Papers of Isaac Newton*. Cambridge (1962).
HARTLEY, SIR HAROLD, The Place of Berzelius in the History of Chemistry. *K. svenska VetenskAkad. Årsb.* 31–50 (1948).
—— Michael Faraday as a Physical Chemist. *Trans. Faraday Soc.* **49**, 473–88 (1953).
—— *Humphry Davy*. London (1966).
HENNEMANN, G. *Naturphilosophie im 19 Jahrhundert*. Freiburg and Munich (1959).
HESSE, M. B. *Forces and Fields*. London (1961).
HIEBERT, E. *Historical Roots of the Principle of Conservation of Energy*. Madison, Wis. (1962).
JACQUES, J. La thèse de doctorat d'Auguste Laurent et la théorie des combinaisons organiques (1836). *Bull. Soc. chim. Fr. Docum.* 31 (1954).
—— Essai bibliographique sur l'œuvre et la correspondance d'Auguste Laurent, précédé d'une note sur Laurent et Gerhardt. *Archs Inst. gr.-duc. Luxem.* N.S. **22**, 11 (1955).
JEFFREYS, A. E. *Michael Faraday; a List of his Lectures and Published Writings*. London (1960).
JORPES, J. E. *Jac. Berzelius*. Stockholm (1966).
KARGON, R. H. *Atomism in England from Hariot to Newton*. Oxford (1966).
KNIGHT, D. M. The Chemical Elements from Davy to Brodie. D. Phil. thesis, Oxford University (1964).
—— The Atomic Theory and the Elements. *Studies in Romanticism* **5**, 185–207 (1966).
—— *Atoms and Elements*. London (1967).
KOYRÉ, A. *Newtonian Studies*. London (1965).

KUHN, T. S. *The Structure of Scientific Revolutions*. Phoenix edn, Chicago (1964).

McFARLAND, T. *Coleridge and the pantheist tradition*. Oxford (1969).

MERZ, J. T. *A history of European thought in the nineteenth century*. 4 vols., repr. Dover, New York (1965).

METZGER, H. *Newton, Stahl, Boerhaave, et la doctrine chimique*. Paris (1930).

—— *Attraction universelle et religion naturelle chez quelques commentateurs anglais de Newton*. Paris (1937).

MULTHAUF, R. P. *The Origins of Chemistry*. London (1966).

PARTINGTON, J. R., *A History of Chemistry*, vols. 2, 3, 4. 4 vols., London (1961–4).

ROYAL SOCIETY, *Catalogue of Scientific Papers, 1800–1900, 1867–1925*. 19 vols.

SCHENK, H. G. *The Mind of the European Romantics*. London (1966).

SCHOFIELD, R. E. *A Scientific Autobiography of Joseph Priestley (1733–1804)*. Cambridge, Mass. and London (1966).

—— *Mechanism and Materialism*. Princeton (1970).

SNELDERS, H. A. De thermochemische onderzoekingen van Hess (1839–1842). *Chem. Tech. Revue* **20**, 397–9 (1965).

STOUGHTON, J. *Worthies of Science*. Religious Tract Society, London (n.d.).

STUMPER, R. La vie et l'œuvre d'un grand chimiste, pionnier de la doctrine atomique: Auguste Laurent, 1807–1853. *Archs Inst. gr.-duc. Luxemb.* N.S. **20**, 47–93 (1951–3).

THACKRAY, A. *Atoms and Powers*. Cambridge, Mass. (1970).

TRENEER, A. *The Mercurial Chemist*. London (1963).

VIRTANEN, R. Marcellin Berthelot, A Study of a Scientist's Public Role. *Univ. Stud. Univ. Neb.* N.S. no. 31 (April 1965).

WEBER, C. A. *Bristols Bedeutung für die englische Romantik und die deutsch-englischen Beziehungen*. Halle (1935).

WHEELER, T. S. and PARTINGTON, J. R. *The Life and Work of William Higgins*. Pergamon Press (1960).

WHYTE, L. L., ed. *Roger Joseph Boscovich*. London (1961).

WILLEY, B. *The Seventeenth-Century Background*. Penguin, London (1964).

WILLIAMS, L. P. Faraday and the Structure of Matter. *Contemp. Phys.* **2**, 93–105 (1960–61).

—— The Physical Sciences in the First Half of the Nineteenth Century. *Hist. Sci.* **1**, 1–15 (1962).

—— Ampère's Electrodynamic Molecular Model. *Contemp. Phys.* **4**, 113–23 (1962).

—— *Michael Faraday*. London (1965).

WILLSHER, A. P. Daubeny and the Development of the Chemistry School in Oxford, 1822–67. Chemistry B.A. Part II thesis, Oxford University, (1961).

NAME INDEX

SUBJECT INDEX

As far as is possible, subjects are organized under the following headings:

Affinity
Atomism
Chemical substances
Chemistry
Force
Matter